Regional Impacts
of Rising
Energy Prices

Regional Impacts
of Rising
Energy Prices

William H. Miernyk
Frank Giarratani
Charles F. Socher

Ballinger Publishing Company • Cambridge, Massachusetts
A Subsidiary of J.B. Lippincott Company

International Standard Book Number: 0−88410−074−X

Library of Congress Catalog Card Number: 77−25075

Printed in the United States of America

Library of Congress Cataloging in Publication Data

Miernyk, William H
  Regional impacts of rising energy prices.

  Bibliography: p. 117
  1. Petroleum products—Prices—United States. 2. United States—States—Economic conditions. 3. Regional economics. I. Giarratani, Frank, joint author. II. Socher, Charles F., joint author. III. Title.
HD9564.M53   338.2'3   77−25075
ISBN 0−88410−074−X

To Nicholas Georgescu-Roegen

Contents

List of Figures

List of Tables

Preface

The research on which the essays in this volume are based was conducted in the Regional Research Institute at West Virginia University over a period of three years. The study was supported in part by the Economic Development Administration of the U.S. Department of Commerce. The material excerpted here is from a longer report submitted to EDA.[1] Because Chapters 2 through 5 were published elsewhere as journal articles, and deal with the same basic theme, there is bound to be some repetition of ideas among them. An effort has been made to eliminate long repetitive passages. We can only apologize for the residual repetition that remains.

It is impossible to acknowledge adequately all of the debts—intellectual and otherwise—incurred while doing the research reported on here. But the contributions of certain individuals must be mentioned. First, the seminal work on energy, as well as the methodological innovations, of Nicholas Georgescu-Roegen provided much of the intellectual stimulus for the papers published here as Chapters 2 through 5.[2] Richard Newcomb, of the College of Mineral and Energy Resources, West Virginia University, was kind enough to comment on early drafts of some of the papers. He should not, however, be

1. Two articles resulting from this study were not included here or in the EDA report because they duplicate material that is covered in the essays that are presented. They are, William H. Miernyk, "Regional Employment Impacts of Rising Energy Prices," *Labor Law Journal* 26 (August 1975): 518–23; and "Decline of the Northern Perimeter," *Society* (May/June 1976), pp. 24–26.

2. See, in particular, his *The Entropy Law and the Economic Process* (Cambridge, Massachusetts: Harvard University Press, 1971); and *Energy and Economic Myths* (New York: Pergamon Press, 1976).

held accountable in any way for errors in the final product. Pat Choate, of the Economic Development Administration, provided support and encouragement throughout the study. His contribution, particularly during the initial stages of this research effort, went far beyond that of arranging for financial support.

It is a pleasure to acknowledge the work of the Regional Research Institute staff. Present and former members who were involved with this project include: Melissa Wolford, James Maddy, Anthony Loviscek, James Cassell, Patricia Henry, and Alan Mierke. Secretarial duties were ably handled by Carla Uphold and Jean Gallaher.

This study would not have been completed without the assistance of those mentioned above, as well as many others who provided data, answered questions, and offered advice and criticism. While cheerfully acknowledging this assistance, however, each of the authors must accept responsibility for his own results and conclusions.

William H. Miernyk
Project Director

Regional Impacts
of Rising
Energy Prices

Chapter 1

Introduction and Summary

William H. Miernyk

When this study was begun in the fall of 1974, the United States had weathered an early stage of the "energy crisis."

It had survived the oil embargo of 1973, and appeared to be adjusting to the dramatically higher prices for fossil fuels which the embargo and other actions by the Organization of Petroleum Exporting Countries (OPEC) had engendered. The basic hypothesis of this study was that the new structure of energy prices, and future energy price increases, would have differential regional impacts in the United States. Consumers in all regions would, of course, pay more for energy. But energy-producing states, it was hypothesized, would benefit, while energy-consuming states would be adversely affected by the new price structure.

At that time there were no data by which the hypothesis could be tested. In fact, we assumed that it might be some time in the early 1980s before clearly defined trends would be discernible to either support or refute the hypothesis. But events have moved faster than we anticipated. While it is still not possible to provide definitive proof that energy-producing states are gaining in economic well-being vis-à-vis energy-consuming states, there are at least straws in the wind which show that this is so. Some of the indicators are presented in Chapters 2 through 4 of this report. Supporting evidence is provided by a number of other studies.[1]

Chapter 2 provides a general overview of the energy problem, and identifies the 20 energy-producing states in the nation. Energy employment is calculated as a percentage of total employment and as a percentage of each state's export base. The latter is a better measure of specialization in energy production than the former.

The consequences of high and rising energy prices for the Northeast, an energy-deficient region, are examined in Chapter 3. The production and consumption of basic energy by region is presented, as well as average regional prices for natural gas. The Northeast's cost disadvantage is clear. The conclusions reached in this chapter are that there will have to be adjustments to the new structure of energy prices, and that the most difficult task might be that of creating a public awareness of the need for such adjustments. The final conclusion is that, while energy-intensive activities are likely to relocate, the Northeast will not become an economic disaster area.

Chapter 4 deals with a slightly more technical matter than those discussed earlier. The most recent United States input-output table (1967) was adjusted for price changes to 1974. The adjusted table was then used to estimate price impacts on intermediate inputs and value added by sector, between 1967 and 1974. Value added in the energy sectors was then imputed to energy-producing states. While there are few surprises in this chapter, it provides reasonable estimates of the windfall gains enjoyed by the energy sectors and by energy-producing states between 1967 and 1974.

Energy prices are discussed in Chapter 5 within the context of the Clark-Fisher development model. A table of total energy input coefficients is included in which four-digit input-output industries are ranked on the basis of total energy consumption. These industries are also identified on the basis of their location orientation. The objective was to identify the kinds of industries that might be attracted to relatively low-cost energy locations.

Chapter 6, by Frank Giarratani, explores the application of an input-output supply model to energy issues. This study highlights the importance of extractive energy sectors in intermediate production; it also identifies supplying sectors that have the potential of restricting energy output. The use of the model to simulate the impacts of alternative energy allocation schemes is discussed and illustrated.

The next chapter, by Giarratani and Charles F. Socher, deals with the pattern of industrial location under rising energy prices. It presents a brief discussion of the classical theory of location. The concept of income potential is then discussed, and state income potentials for 1970 and 1974 are calculated. On the basis of these calculations, as well as coefficients of localization that were calculated for the same two years, the authors concluded that "manufacturing is predominantly market-oriented and apparently becoming more so." They also conclude that energy-producing states will become increasingly important markets, and thus should attract more manufacturing activity.

The essays in this report make a distinct break with the traditional literature of regional economic growth and development. Regional growth studies, like their national counterparts, have been demand-oriented.[2] And regional development policy, both in the United States and abroad, has been concerned with demand deficiencies at the regional level.

This volume, however, deals with supply constraints and the effects of these constraints on the location of industry and regional development. This shift in emphasis is not the result of whim or caprice. It is an attempt to keep up with actual changes, both physical and economic, in the world around us. As Chapter 2 points out, energy has not been thought of as an important locational determinant in the past, except in a few capital-intensive electrolytic processes. But it is likely to be one of the more important locational determinants for a fairly wide range of manufacturing activities in the future. And because changing energy costs could have an important influence on demographic trends, they could have even broader impacts on levels of regional activity and regional growth rates. This is becoming increasingly well-known in policy-making circles, although the issue has been rather widely ignored by academic economists and regional scientists.[3]

If one makes the assumption that resources, including energy, are virtually unlimited—as most growth theorists have done, either implicitly or explicitly—limits to economic growth can be ignored. The father of modern macroeconomics, John Maynard Keynes, wrote that the day would come when our material wants would be so thoroughly sated there would be no need for any further accumulation of wealth.[4] Under these circumstances, the direction of regional policy would be clear. Demand stimuli, coupled with investment in human resources, would be prescribed for lagging regions.

Regional problems become far more complex, however, when supply constraints are considered, and a growing number of economists now recognize that such constraints exist.[5] Supply constraints can impose limits to growth; and because of the uneven geographic distribution of resources, they will have differential regional impacts. In a market economy, resource constraints will cause some regions to grow and will lead to decline in others.

It is far more difficult to reach a national concensus on regional policy now, when supply constraints are explicitly recognized, than it was in the early 1960s, when the belief was widespread that the secret of unrestrained growth without inflation had been found. We hope, however, that this volume will contribute to rational debate of regional issues, and that it will be useful to those faced with the difficult task of formulating a revised regional policy to deal with the new regional problems the nation now faces.

NOTES TO CHAPTER 1

1. See, for example, Hillard G. Huntington and James R. Kahn, "Industrial Location and Regional Energy Prices," Federal Energy Administration, Office of Economic Impact Analysis, Working Paper 76—WPA—35 (October 1976); David Puryear and Roy Bahl, "Economic Problems of a Mature Economy," Metropolitan Studies Program, Maxwell School of Citizenship and Public Affairs, Syracuse University, Occasional Paper No. 27 (April 1976); Roy Bahl, Alan Campbell, David Greytak, and David Puryear, "Public Policy Implications of Regional Shifts in Economic Activity," prepared for the Northeast Congressional Coalition, Metropolitan Studies Program, Maxwell School (October 18, 1976); and William H. Miernyk, "The Changing Structure of the Southern Economy," Southern Growth Policies Board, Occasional Paper No. 2 (January 1977). There is also a vast literature on the shift of industry from the Snowbelt to the Sunbelt. While the availability of energy at relatively low prices is only one of the causes of this shift, it is probably one of the major causes. For a summary discussion of the latest episode in the migration of industry in the United States, and the controversy surrounding it, see Gurney Breckenfeld, "Business Loves the Sunbelt (and Vice Versa)," *Fortune*, June 1977, pp. 132—46.

2. See, for example, Harry W. Richardson, *Regional Growth Theory* (London: Macmillan, 1973), especially the extensive bibliography on pp. 237—53. See also William H. Miernyk, "Some Regional and Environmental Consequences of Rising Energy Prices," paper presented to the Second Tokyo Environmental Conference, August 19—20, 1976.

3. For an excellent summary of the organizations that have been involved in the discussion of contemporary regional issues, see Robert W. Rafuse, Jr., *The New Regional Debate: A National Overview* (Washington, D.C.: National Governors' Conference, Center for Policy Research and Analysis, April 1977).

4. See "Economic Possibilities for Our Grandchildren," in *Essays in Persuasion* (New York: Norton, 1963), pp. 358—73.

5. For pioneering work along these lines, see Nicholas Georgescu-Roegen, *The Entropy Law and the Economic Process* (Cambridge, Massachusetts: Harvard University Press, 1971); and *Energy and Economic Myths* (New York: Pergamon Press, 1976). The latter is a collection of essays originally published between 1952 and 1972.

Chapter 2

Regional Economic Consequences of High Energy Prices in the United States

William H. Miernyk

INTRODUCTION

Except in isolated cases, low-cost energy has not been considered to be an important determinant of industrial location. Aluminum reduction plants, and a few other capital-intensive electrolytic processes, have always located where electric energy prices were low. But conventional location theory postulated that managers would try to minimize transport costs when choosing plant locations. If further investigation showed that some plants should be located elsewhere, because of low labor cost or cheap energy, such locations were considered to be "deviations" from the minimum transport-cost point.

The low priority given to energy as a locational determinant made sense at a time when energy was cheap. When it cost little to produce energy it also cost little to transport it. Thus, most industrial plants could be attracted to sources of raw materials, markets, urban nodes or where labor costs were low. Energy could and would be brought to them.

Why were energy prices low? Since prices are determined by sup-

Reprinted by permission from *The Journal of Energy and Development*, Volume 1, No. 2 (Spring 1976): 213–39.

This article was presented in preliminary draft form to a Seminar on Economic Development, held August 13–14, 1975, in Washington, D.C. under the sponsorship of the Economic Development Administration of the U.S. Department of Commerce and the Hudson Institute. The author wishes to acknowledge the interest, support, criticism, and/or research assistance of Pat Choate, Edward K. Smith, Victor A. Hausner, Nicholas Georgescu-Roegen, Richard Newcomb, James Maddy, Melissa Wolford, and James Cassell.

ply and demand, it must have been because the supply of energy was abundant relative to demand. Competition among producers of oil and coal kept prices down. Since final consumers would have been willing to pay higher prices than those actually charged, they enjoyed a "consumers' surplus." Meanwhile, industrial and commercial purchasers of energy were collecting economic rents on the low-cost oil, coal or electric energy they purchased as intermediate goods. The present discussion focuses primarily on coal as a source of cheap energy, but much of what is said would apply to oil, particularly in the international market.

Throughout most of its history, the American coal industry has been highly competitive. Members of the industry typically speak of coals rather than coal, because of wide variations in heat, water, and sulfur content, as well as in hardness. But within each class, coal may be viewed as a homogeneous product. Thus, it is sold on the basis of price rather than product differentiation.

In addition to *intra*industry competition, there was always competition from other sources of energy, particularly oil. Indeed, it was the threat of oil which led John L. Lewis to negotiate his famed "mechanization agreement" of 1950. Lewis was able to convince the operators that their survival depended upon mechanization and modernization, and he did this knowing that thousands of miners would be displaced as a result of the rationalization of the industry.[1] Miners had long been accustomed to stretches of cyclical unemployment. But following the 1950 agreement, there was a great deal of chronic, structural unemployment among displaced coal miners. Those who were lucky enough to hold their jobs were paid more than they had been in the past, however, since new and higher skills were required.

The mechanization agreement gave the American coal industry a new lease on life although continued competition from oil kept prices low. Between 1950 and 1955, coal prices dropped from $4.84 to $4.50 per ton, or 7 percent, while other wholesale prices rose 7 percent and the consumer price index went up more than 11 percent. The drop in coal prices would have been larger except for an accompanying cut of 10 percent in production.[2]

Low energy prices were a boon to industrial and urban America. Cheap energy made an important contribution to the urbanization (and suburbanization) of the nation.[3] Low-cost energy permitted the adoption of innovations in agriculture which greatly reduced the demand for labor while permitting massive annual increases in output in this sector. Cheap energy also contributed to mechanization and automation in America's mass-producing industries, and to the relative price stability of the entire economy in the 1950s and early

1960s. Meanwhile, downward pressure on coal prices exerted downward pressure on coal wages. The wages of highly skilled miners began to compare favorably with those of industrial workers, but mechanization was not accomplished overnight, and average wages in coal mining lagged behind the industrial wage average. Although they are declining in number, and contribute only a fraction to the coal industry's total output, "dog-holes"—small nonmechanized miners—even today have not disappeared from the hills of Appalachia.

Throughout much of the nation's history, energy-producing regions have "subsidized" the growth of urban America by providing an abundant supply of energy at low and stable prices. Meanwhile, relatively low wages and widespread unemployment added to the economic distress which appeared to be endemic to rural Appalachia and other coal-producing regions. When urban and industrial America moved into the Age of Affluence, partly because of cheap energy, the coal-producing regions remained poor. But the late 1960s marked the end of the era of cheap energy. Future historians will regard the first half of the present decade as a turning point or watershed in the history of the industrial nations of the world. Up to the present, industrial society has not yet recovered from the events that led to the "energy crisis" of 1973. Nor is it evident that most political leaders are willing to acknowledge that the recent escalation of energy prices is anything more than a transitory phenomenon.

This paper is based on the assumption that high energy prices are here to stay, and that future demand-supply relationships are likely to keep energy prices rising faster than other wholesale prices. It is concerned primarily with the *regional* consequences of the changing relationship between energy and other prices, but before these are discussed the causes of the "energy crisis" are considered. The paper is largely exploratory. It outlines an area of research which some of my associates and I plan to investigate in depth, and presents some hypotheses which we plan to test. It concludes with a discussion of some implications for regional policy if events show that the basic assumption stated above is correct.

WHAT HAS HAPPENED TO ENERGY PRICES, AND WHY?

The energy sector that has attracted most public attention is petroleum. As we will see later, perhaps more attention should have been paid to what has happened to coal prices. Relative to the prices of other commodities, petroleum prices held steady during the 1960s, and declined between 1966 and 1973. The world's attention was riveted on petroleum prices in September 1973 when the Arab nations reduced production and imposed an embargo against most of their major customers, including the United States. The price of oil quadrupled, and in the year that followed the ratio of petroleum to other commodity prices almost doubled.

The reaction to the oil embargo in the United States was one of outrage. A number of eminent economists labeled the embargo as a form of blackmail, and urged the nation's leaders not to waver in their support of Israel both during and after the October 1973 Middle East war. There is no doubt that the 1973 embargo was designed to strengthen the hand of the Arab nations in the Middle East. But the decision to cut production and raise prices was the culmination of a chain of events that goes back much farther than the October war.

In 1959, seven major multinational oil companies agreed to cut the price they were paying to oil-producing nations. This led five of the oil-producing nations—Saudi Arabia, Iran, Iraq, Kuwait, and Venezuela—to meet in Baghdad in 1960 to form the Organization of Petroleum Exporting Countries (OPEC). As John Goshko pointed out in a *Washington Post* story, the new organization was greeted rather scornfully by diplomats and Western businessmen as "a quarreling collection of camel sheikhdoms and banana republics."[4]

Other oil-producing countries began to join the organization which was originally conceived by Juan Pablo Perez Alfonso, a Venezuelan oil expert generally acknowledged as the "father of OPEC."[5] The new members included Algeria, Ecuador, Gabon, Indonesia, Libya, Nigeria, Qatar, and the United Arab Emirates.

By 1973, OPEC was clearly in a position to control petroleum output and thus, petroleum prices. The rest of the story is well known. Oil prices have been cut back slightly from their peak, but remain approximately four times as high as the pre-embargo levels. In recent years, while other pejorative terms have been used to describe the OPEC nations, there have been few references to "camel sheikhdoms and banana republics." In the United States, where life appears to center as much around the automobile as it does around the home, there is a pitiable yearning for the "good old days" of cheap gasoline. But despite the unwillingness of many political leaders to face the issue squarely, the return of cheap gasoline is about as likely as the return of the Model—T Ford.

Until the late 1960s, all forms of energy were bargains to the American consumer. The price of electricity fell throughout the 1960s, relative to other prices. Coal, gas, and petroleum prices were either stable or declining. Indeed, U.S. petroleum prices had been declining, relative to other prices, for nearly two years before 1959 when the multinational companies issued their price-cutting edict to the oil-producing nations.

Coal prices had started to rise in the United States well before the 1973 oil embargo and the subsequent concern about the "energy crisis." Until 1969, coal prices had been increasing no faster than other wholesale prices, but between 1969 and 1970 the ratio of coal to other wholesale prices rose sharply. The average price of coal— which is a composite of open-market and long-term contract prices— increased from $4.99 to $6.26, a jump of 25 percent. There was little change in relative coal prices until 1973. But when petroleum prices zoomed upward following the Arab embargo, coal prices made another mighty leap upward. By the end of 1974, the average price of coal had reached an all-time peak, and was about three times as high as the pre-1969 level.

Figure 2-1 shows the relationship of open-market coal prices to other energy prices, and to wholesale commodity prices. Coal prices are at the top of the heap. The lower portion of the figure shows the recent historical relationships between gas, petroleum, and coal price indexes. Both the gas/coal and the petroleum/coal ratios were declining between 1966 and 1973. The petroleum/coal ratio pulled up sharply in 1973 as petroleum prices temporarily pulled away from coal prices. The figure shows, however, that this historical relationship is rapidly being reestablished. It is not likely that coal prices will continue to rise more rapidly than oil prices; it is likely that the rough "parity" represented by the period 1962-1967 will be maintained in the future. If the price of imported oil were to fall and return to, say, its 1970 level—with appropriate adjustments for inflation—coal prices would be almost certain to drop. If this were to happen, we would be back in the era of cheap energy. But is it likely to happen?

WHAT WILL HAPPEN TO ENERGY PRICES, AND WHY?

There has been a virtual explosion of books and articles dealing with petroleum prices since the 1973 embargo. Interestingly, much of this literature is highly optimistic about the return of cheap oil. Many political leaders also implicitly accept the notion that present high energy prices are temporary. This must be the case, since there has been a clear-cut unwillingness on the part of Congress to initiate an

Figure 2–1. Ratio of Wholesale Fuel Prices to All Commodities

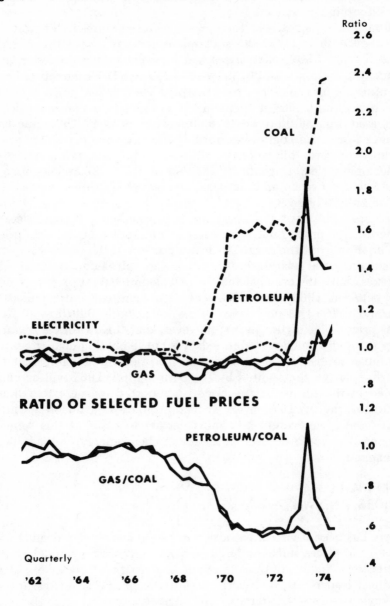

Source: *Cleveland Trust Business Bulletin*, April 1975.

energy policy directed toward restraints on consumption. Judging by their public pronouncements, some political leaders appear to believe that lower energy prices are not only possible but likely.

One view has it that the "energy crisis" is essentially political in origin, and has little or nothing to do with supply-demand relationships. A leading exponent of this view is M.A. Adelman, who "argues that OPEC's recent successes should be attributed in large part to misguided policies of the U.S. State Department."[6] In a widely quoted article Adelman has said: "Without active support from the United States, OPEC might never have achieved much."[7] The trouble started, according to Adelman, when Syria blocked the trans-Arabian pipeline in May 1970 to obtain higher transit-rights payments, while Libya enforced production cutbacks and higher taxes. Other OPEC countries then imposed similar tax increases.

Adelman believes that the United States missed a tremendous opportunity in May 1970. He feels that the government could have "convened the oil companies to work out an insurance scheme whereby any single company forced to shut down would have crude oil supplied by the others. . . ."[8] What Adelman has suggested is the formation of a government-sanctioned cartel to provide countervailing power to the OPEC cartel. He feels that at this stage of its development OPEC was "unprepared for conflict"—presumably he meant economic conflict in this context. But the turning point came, according to Adelman, when the United States convened a meeting in Paris of the OECD nations (the countries belonging to the Organization for Economic Cooperation and Development). At this meeting, Adelman believes, OECD could have outbluffed OPEC. Instead, an OECD spokesman praised the international oil companies for proposing higher and escalating taxes. This same spokesman also stated that OECD "had not discussed 'contingency arrangements for coping with an oil shortage.' This was an advance capitulation."[9] Adelman does not suggest what these contingency arrangements might have been. He also claims that the United States Undersecretary of State "told the Shah of Iran the damage that would be done to Europe and Japan if oil supplies were cut off. Perhaps this is why the Shah soon thereafter made the first threat to cut off supply."[10] Such statements corroborate Goshko's view that before 1973 OPEC was viewed as "a quarreling collection of camel sheikhdoms and banana republics."

In his analysis of the causes of the energy crisis, Adelman concentrates on political matters and does not discuss such issues as changing demand-supply relationships. He also evidently continues to believe that the OPEC cartel is inherently unstable. In a letter to the editor of the *New York Times* (June 18, 1975), he stated his conviction that the drop in oil production beginning in 1973 has resulted "in excess capacity [that] has made the cartel highly vulnerable."

Adelman's view has been echoed by Robert Z. Aliber (*Wall Street Journal*, March 20, 1975). Aliber asserted that "the oil cartel is in the early stages of a breakdown." He predicted that "in the next several months, the demand for OPEC-produced petroleum will decline sharply," and claimed that "OPEC exports have declined even more rapidly than the demand in the consuming countries." This, he says, was the result of two mild winters in succession and also because "1975 autos consume substantially less gasoline than the late 1960 models now being junked."

Aliber's highly optimistic views followed a rash of earlier press releases, given prominent attention by the *Wall Street Journal*, that some OPEC nations were cutting prices. For example, the *New York Times* reported, on March 2, 1975, that Abu Dhabi had cut the price of its oil by $.55 a barrel. This, it turned out, had resulted from an earlier decision by OPEC to allow Abu Dhabi and "one or two other smaller OPEC members" to lower the *price premiums* they had been allowed to charge because of the relatively high quality of their low-sulfur crude oil (*Wall Street Journal*, February 27, 1975). This modest price adjustment, in conjunction with reported cutbacks in production, led a number of journalists to wax enthusiastically about the "impending breakup of [the] OPEC cartel"—the title of Aliber's *Wall Street Journal* article.

Writing in the *New Republic* (February 15, 1975), Melville Ulmer drew on an earlier paper by John Blair, former chief economist for the Senate Antimonopoly Subcommittee, and suggested that the real culprit in the energy crisis was the *domestic* oil cartel: ". . . The American public," he wrote, "is caught in a pincer movement between a domestic cartel and a foreign cartel, essentially two arms of the same body." He stated his belief that "if the power of the federal government were used effectively . . . the domestic cartel could be broken almost at once, and the stranglehold of the OPEC countries severed within two years."

Ulmer gave no basis for his rather precise time estimate. But the way to destroy the two cartels, he asserted, is to assign each domestic company a production quota, and to impose tax penalties, on a per barrel basis, for failure to meet these quotas. This scheme totally ignores limitations on domestic petroleum supplies; it implicitly assumes a completely elastic supply schedule for oil in this country. Ulmer's whimsical proposal was topped by Charles Bluhdorn, chairman of the conglomerate known as Gulf and Western, who proposed that an antitrust suit be launched against OPEC "for conspiring to fix oil prices" (*Wall Street Journal*, May 20, 1975). The fact that a reputable newspaper would report such a nugatory proposal, and actually lend it editorial support (*Wall Street Journal*, August 28, 1975), is indicative of the near-panic which was engendered in some quarters by the "energy crisis."

Throughout the first quarter of 1975, the *Wall Street Journal* continued to give prominent coverage to news stories suggesting "the impending" breakup of the OPEC cartel. Other business-oriented publications joined the chorus. *Fortune* carried a long article by Louis Kraar which recalled earlier references to "camel sheikhdoms and banana republics." Kraar asserted that:

> Politics, not economics, dominates the organization's decision making. In fact, no thorough economic analysis had undergirded any of OPEC's daring moves. Neither the organization nor its members are equipped to perform sophisticated economic studies.[11]

Perhaps it was "analyses" such as these which led David Gordon Wilson to write in the *New York Times* of February 23, 1975, that: "The Flat-Earth Society has been reborn. It is now the Cheap-Energy Society. It seems that most politicians, economists, and news commentators are enthusiastic members." He went on to say:

> The fervor with which they are denouncing the President's proposals to raise energy prices will, along with past beliefs in the divine right of kings, the sanctity of child labor and the inferiority of women, be puzzled over by future historians.

How good were the forecasts made by Aliber, Kraar, and other putative members of Wilson's Cheap-Energy Society? The *New York Times* of May 25, 1975 reported that Arab oil accounted for 22.7 percent of U.S. first-quarter imports in 1975, compared with 17.9 percent in the fourth quarter of the previous year. Saudi Arabia's share had increased from 10.6 percent to 13.4 percent of the total. On his visit to the United States in May 1975, the Shah of Iran predicted that there would be an *increase* in oil prices following the end of OPEC's voluntary six-month price freeze in September. In a full-page advertisement, carried by the *Wall Street Journal* (June 6, 1975), the U.S. representative of the National Iranian Oil Company claimed that the prices of exports from OCED countries to OPEC members had increased by 25 percent during 1974, and were expected to increase another 10 to 15 percent by the end of September 1975. He contended that oil-exporting nations had thus lost between 30 and

35 percent of the purchasing power of their dollar earnings from oil exports between January 1974 and September 1975. The Shah of Iran also used these estimates, but concluded that oil prices would rise less than the 30 to 35 percent increase in OPEC's import prices. There was growing evidence that the Shah's forecast was being taken seriously.[12]

Not all analysts had agreed with the Aliber-Kraar forecast. Anthony Parisi, for example, had stated that "the present surplus of oil is strictly a short-term phenomenon steming mainly from inertia."[13] OPEC, he noted, had been cutting back output to take up slack in demand, but it had not been able to do this fast enough. Parisi felt, however, that there was no reason to assume that OPEC cannot make further cuts when needed. In the long term, he stated "the cartel's grip on the oil tap will weaken—unless it raises prices now."[14]

Parisi is one of the few energy analysts who discussed the current petroleum situation in terms of both demand *and supply*. He cited estimates by Arnold Safer, an Irving Trust Company economist, that non-OPEC oil supplies could increase from 16 million barrels daily in 1974 to 23 million barrels daily in 1977. By 1977, however, Safer has estimated Free World consumption at 47.5 million barrels a day. If both these supply and demand projections are accurate, OPEC could be caught with a "true oversupply" of one-half million barrels a day, and these are the conditions, says Parisi, that start price wars. Parisi concluded, however, that "it is in OPEC's best interests to raise the price of oil now." If nct, inflation would erode OPEC's revenue base. By maximizing income while it is in a position to do so, Parisi feels the cartel could build up a sufficient financial reserve to allow it to curtail production still more in 1977, if it becomes necessary to do so to maintain prices.*

For reasons that are difficult to fathom, most economic analyses of the energy situation—particularly with respect to petroleum—have largely ignored the supply side of the equation. Economists and journalists of the Adelman-Aliber-Kraar persuasion appear to believe that present high energy prices are a departure from some sort of cheap energy norm. And many journalists appear to be willing to accept, without reservation, Disney's First Law—Wishing Will Make It So. There have been exceptions, of course, such as Anthony Parisi cited earlier. Another notable exception is the distinguished Vanderbilt economist, Nicholas Georgescu-Roegen. In a series of books and articles published over the past decade, Georgescu has stressed the

---

*In spite of a rash of newspaper stories about a potential glut in oil production, the glut did not materialize. OPEC raised prices an additional 10 percent in January 1977 (note added October 1977).

importance of the resource base to sound economic analysis—something that few economists have done since the time of Thomas Robert Malthus. He also has pointed out repeatedly that mankind has available a declining stock of *free* (as opposed to bound) energy.[15]

Conventional economics, Georgescu asserts, "portrays the economic process as self-sustaining, circular flow between 'production flows' and consumption flows.' "[16] This is a serious methodological error, Georgescu believes, because the economic process "alters the environment in a cumulative way. . . ."[17]

One of the myths Georgescu has criticized is the assertion that if zero population growth (ZPG) could be achieved, mankind would "no longer have to worry about the scarcity of resources or about pollution. . . ."[18] Another economic myth, in this economist's view, is that man will forever find new sources of energy, and new ways of harnessing that energy. "The idea is that, if the individual man is mortal, at least the human species is immortal."[19]

Georgescu has been unique among modern economists in his efforts to relate economics and physics, particularly thermodynamics. He has tried to bring to the attention of economists certain incontrovertible conclusions of thermodynamics. He has pointed out, for example, that "available energy is continuously transformed into unavailable energy . . . [because] all kinds of energy are gradually transformed into heat and heat becomes so dissipated in the end that man can no longer use it."[20] But Georgescu is no Cassandra who believes that mankind is rushing pell mell to early extinction. Despite the abstract, and sometimes esoteric, nature of his analysis, he arrives at a series of practical conclusions. Although highly idealistic, few of these conclusions would be rejected by the average individual who professes to have a social conscience. He argues, for example, that "until either the direct use of solar energy becomes a general convenience or controlled fusion is achieved, all waste of energy . . . should be carefully avoided, and if necessary, strictly regulated."[21] He also would like to eliminate "extravagant gadgetry" and the planned obsolescence of clothing, furniture, autos, and indeed, most other goods.

Strangely, however, energy economists and others concerned with the "energy crisis" have paid little attention to Georgescu's important work. Several of the prescribed "cures" for the "energy crisis" completely ignore the supply side of the problem. The oversimplified "political" analyses of energy prices assume unlimited supplies of *domestic* energy sources, despite evidence to the contrary. Reputable scientific organizations have warned that the energy situation is going to get worse, not better. Twice within the past two years, for example, the United States Geological Survey has "sharply lowered its estimates of how much oil and natural gas in the U.S. remains to be discovered," according to a *Wall Street Journal* report of May 8, 1975. The National Academy of Sciences and the United States Senate National Ocean Policy Study estimate that U.S. oil and gas resources could be exhausted by the end of the century (*Washington Post*, May 11, 1975).[22] Commenting on the views of M. King Hubbert, a past president of the Geological Society of America who has worked for both private industry and the United States Geological Survey, an article in *Science* stated that he "believes the petroleum era will be a brief blip in human history."[23]

THE LONG-RUN ENERGY OUTLOOK

The publication a few years ago of *The Limits to Growth* started a controversy about the future availability of resources and energy that is far from settled.[24] *The Limits to Growth* was a popularized version of a report supported by the Club of Rome which utilized Jay Forrester's method of "Systems Dynamics," a computer simulation technique. The study was criticized for its allegedly unscientific character by, among others, such luminaries in economics as Kaysen and Solow.[25]

Most economists, it appears, are technological optimists. This is true, of course, of many physical scientists as well. Those who see the breeder reacter as the *deus ex machina* of the energy crisis fall into this category. Some have stated that breeder reactors will produce more fuel than they consume. Georgescu calls this "the fallacy of entropy bootlegging." He says:

> ... the stark truth is that the breeder is in no way different from a plant which produces hammers with the aid of some hammers ... even in breeding chickens, a greater amount of low entropy is consumed than is contained in the product.[26]

The answer of the technological optimists to the energy crisis is: Not to worry! Some economists, such as Nordhaus, whose work is cited with approbation by Solow in his Ely lecture, feel that we will have nuclear fusion—and virtually unlimited energy—before we use the last drop of oil. If developed, nuclear fusion would indeed assure a long-term supply of electrical energy. But nothing is said about the ways in which this energy could be used for the propulsion of vehicles. Presumably, the optimists feel that if engineers and scientists are clever enough to develop the fusion process, some of them will be clever enough to devise economical ways to use this new source of energy to move cars, trains, and airplanes. Perhaps one might be forgiven, however, for wondering if this is not another instance of Disney's First Law at work.

The technological optimists—whether economists or physical scientists—pay little attention to coal as a source of future energy. In his book dealing with the alleged failure of U.S. energy policy, for example, Mancke devotes only three pages to coal. Much of his discussion is based on a quotation from an unpublished paper by R.L. Gordon, which pithily describes the shortcomings of coal. Mancke concludes that:

> Unless it is first converted into synthetic crude oil or natural gas, coal cannot be burned in internal combustion engines, is highly inefficient for space heating, and is too "dirty" to be used in most industrial processes. Because of these well-known deficiencies, coal's share of our total energy supply has been falling for over 40 years.[27]

While writing this book, which was done in obvious haste, Mancke evidently did not take the time to examine projections of energy supply. All projections worthy of serious attention have indicated that coal will provide a *growing* share of the U.S. energy supply for the remainder of this century.[28] He also did not trouble to look into the present status and future prospects of such technological prospects as coal gasification.[29] In his Ely lecture, Solow discounted the prospects of coal liquefaction and gasification. He ventured the opinion that oil derived from shale was a far more likely possibility. This assertion is not verified by a perusal of present plans for federal support of new energy programs.

As part of its response to the "energy crisis," the U.S. government has reconstituted its energy research program. The Atomic Energy Commission, with its narrow focus on nuclear energy, was replaced by the Energy Research and Development Administration (ERDA). Unlike its predecessor, ERDA is not putting all its energy eggs in one long-run basket. Recently, for example, the government cut back research on the breeder reactor. The project will not be dropped, but its time horizon has been extended. For the immediate future, ERDA has scheduled a stepped-up program of research on the improved use and conservation of fossil fuels. This will include research on both high- and low-Btu coal gasification and on various attempts at coal liquefaction. In spite of Solow's guess that oil from shale would come first, it is not even considered as an early contender at present.[30]

There have been a number of forecasts of the future pattern of energy production in the United States.[31] Most show moderate to substantial increases in coal production. The *Project Independence Report* of the Federal Energy Administration has two coal projections with regional breakdowns which are particularly useful for present purposes. The two projections have been dubbed the "business-as-usual" (BAU) and the "accelerated development" (AD) forecasts. The first projection is no doubt on the low side, and the second seems high. A good compromise is the midpoint of these two, which would yield an annual production of 1.6 billion tons of coal by 1985. This compares with about 600 million tons produced in 1974. The most rapid *rate* of increase in coal production, because of the low base from which it is measured, will be in the West. But the largest amount of coal will still be produced in Appalachia by 1985.

Figure 2-2 shows the Project Independence coal regions. These regions include parts or all of Appalachia, Southern Illinois, the Ozarks, the Upper Midwest, and the Four Corners—all of which have been classified in the recent past as depressed areas.

Figure 2-3 shows a set of regional coal supply curves in 1985 based on the "business-as-usual" assumption. These probably should not be taken too seriously as actual supply schedules, but they illustrate quite clearly the relative importance of each of the coal-producing regions a decade hence.

The supply schedules show interesting contrasts. Much of Appalachia's coal production will come from existing and new deep mines (segments B and D on the supply schedules), while all of the coal in the Deep South and the three western regions is projected to come from surface mines. The Midwest, which is already a major coal-producing region, ranking just behind Appalachia, shows a mixture of new deep and surface mine capacity but a larger proportionate increase in surface mining.[32]

Both the Federal Energy Administration and ERDA foresee a substantial increase in coal capacity during the next decade. In view of the nation's dwindling oil and gas reserves, this appears to be the only way that the United States can become *less* dependent upon foreign energy sources in the foreseeable future. To speak of *total* energy independence is to stretch one's credibility beyond the breaking point.

What will be the economic consequences of the revival and expansion of coal? First, there are some consequences that one can already observe. Because coal revenues have increased more rapidly than costs of production—including labor costs—coal operators can earn huge profits. A substantial part of these profits, especially those earned by the larger energy corporations, are being plowed back into

Figure 2–2. Project Independence Coal Regions

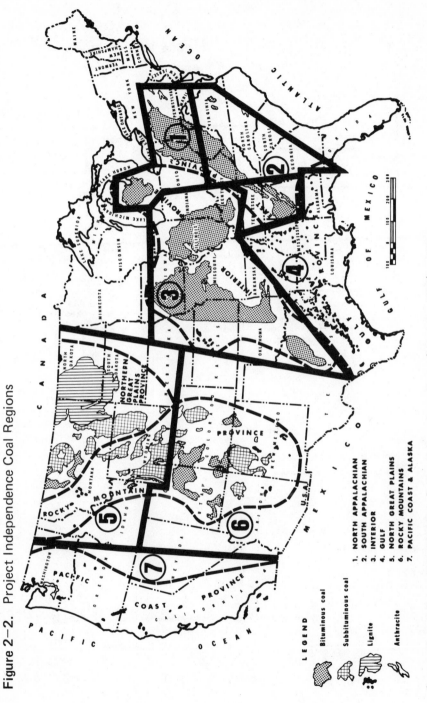

Source: Federal Energy Administration, *Project Independence Report* (November 1974), p. 107.

Figure 2–3.   1985 Business-as-Usual Regional Supply Curves

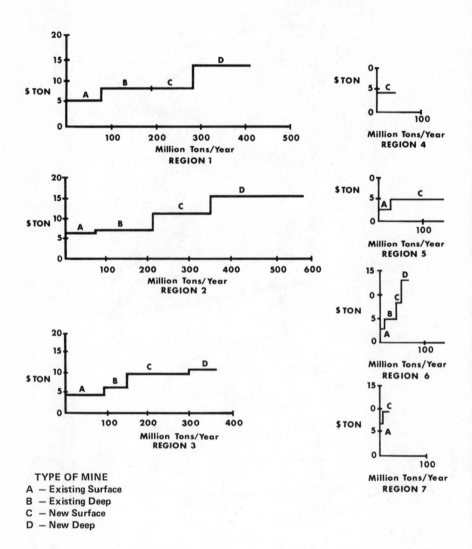

TYPE OF MINE
A — Existing Surface
B — Existing Deep
C — New Surface
D — New Deep

Source: Federal Energy Administration, *Project Independence Report* (November 1974), p. 102.

new capacity. In Appalachia, much of this investment is going into new deep mines at somewhere between $20 million and $30 million per mine.

Even during the present recession, the important coal-producing states have been riding high. They have felt the impact of declining manufacturing employment; but with expanding coal production, employment in mining has been rising. Most important, above-average increases in coal prices have brought huge increases in coal revenues to some of these states. They have reported budget surpluses; their unemployment rates compare favorably with national averages, and their unemployment compensation funds are in good shape. Personal income in these states is rising faster than in the nation as a whole, and in some cases there has been a reverse flow of migrants. West Virginia, for example, had been losing population for at least twenty years before 1970. It is now gaining. Intercensual estimates by the Bureau of the Census show that southern Appalachia—the most seriously depressed subregion of Appalachia during the 1950s and 1960s —is now gaining people. Boone County, West Virginia, has gained between 10,000 and 15,000 inhabitants, or 40 to 60 percent of its population, since 1970. Logan County, another major coal producer in the region, is bustling with return migrants, many from such auto centers as Detroit or Lordstown, Ohio (*New York Times*, May 4, 1975). Both of these counties were long included among the nation's most intractable depressed areas.

Other economic indicators, such as rates of increase in personal income and unemployment rates, have been turned around by the coal revival. In relative terms, and in some cases in absolute terms, these indicators show the energy-producing (EP) areas of the nation to be better off than the country as a whole. Meanwhile, the energy-consuming (EC) states—which include some of the nation's major industrial producers—have been much more seriously affected by the recession than the energy producers.

The immediate regional impact of high energy prices has been to cushion the recession in EP states, and to worsen it in EC states. This is not to suggest that economic conditions in coal-producing states are currently *better* than those in the nation as a whole. The economic consequences of decades of economic deprivation are not wiped out by a few years of relative prosperity. One would have to be an avid believer in the trickle-down hypothesis to argue that the benefits of coal prosperity are being diffused to all parts of the coal-producing regions. There has been some diffusion of economic benefits because of the indirect and induced effects of rising coal revenues and wages. Some has also taken place as a result of increased tax collections on coal. Regional aggregates tell us nothing, however, about the actual distribution of benefits.

The problem of distribution is an important one which deserves further careful study. It is a separate issue, however, and cannot be given the space it deserves in a paper which has a *regional* development focus. But one can say that as coal revenues and tax receipts continue to rise, the coal-producing states will be in a better position than they have been in the past to provide benefits to the residents of the hills and hollows who have always been complete strangers to the Affluent Society.

THE LONG-RUN REGIONAL CONSEQUENCES OF HIGH ENERGY PRICES

The preceding section discussed some of the observable economic impacts resulting from high energy prices, particularly coal prices.[33] In this section we will consider some of the long-run consequences of high and rising energy prices. This will necessarily involve more speculation than the earlier discussion.

If the long-term price of coal increases, as it is likely to do, there will be a regional shift of real income from energy consumers to energy producers. Since energy prices are likely to rise faster than the prices of other primary and intermediate inputs, the "terms of trade" between EPs and ECs will shift in favor of the former. Factors engaged in the production of energy will be rewarded more generously than they have been in the past. And energy-producing states will impose higher taxes on coal to capture some of the economic rents formerly going to consumers in other regions.

Table 2–1 shows the nation's leading energy-producing states—those likely to gain as the result of high and rising energy prices. Column 9 shows basic energy employment as a percent of total non-agricultural employment, and Column 10 shows energy employment as a percent of "export-base employment." Basic energy employment is defined here to include employment in the production of coal, oil, and natural gas. Export-base employment is generally defined as employment in selected agricultural, mining, and manufacturing activities, since typically a substantial part of the output of these sectors is intended for shipment outside the state of production.

We were unable to develop estimates of agricultural export-base employment by state, so the figures in Column 3 relate to mining and manufacturing only. Not all of the output of these two sectors is for export, but no effort was made to adjust for the amount of employment devoted to production for in-state consumption. The hope is that this will offset our failure to include agricultural employment in the export base. The definition of energy employment used here is a narrow one. It does not include, for example, employment in electric utilities; and some states, such as Ohio, Pennsylvania, and West Virginia, are substantial exporters of electric energy.

Only three of the EPs listed in Table 2—1—Alaska, Colorado, and Kansas—had above-average per capita income in 1973. Per capita income in states such as West Virginia, Louisiana, New Mexico, and Kentucky ranged from 77 to 81 percent of the national average. Because they are starting from relatively low income bases, and are also heavily dependent upon energy production relative to their export bases, these are the states most likely to show rapid increases in per capita income during the years ahead. The coal-producing regions of Appalachia and the Far West could be transformed into relatively prosperous areas. Meanwhile, parts of some of the nation's most prosperous states—such as Michigan and Connecticut—could become chronically depressed areas.

Regional shifts in real income from ECs to EPs could be brought about in a number of ways. The first impact, already well under way, will result from rapid increases in the price of electric energy. Early in this decade—before the recent substantial price increases in coal and oil—the Federal Power Commission projected that electric bills would triple by 1990 (*Wall Street Journal*, April 17, 1972). If this projection is adjusted only for the price increases that have occurred since the fall of 1973, one would conclude that there will be a six or sevenfold increase in electric bills by 1990. And this adjustment does not take into account *further* price increases between now and the target year.

Historically, electric energy has been one of the greatest bargains available to American consumers. Even during periods of general price increases, the *relative* price of electric energy has declined. Whether electricity is generated from coal or oil in the future, however, its price is likely to rise relative to other prices. Technological optimists will argue that this will only hasten the day when most electricity will be generated by nuclear rather than fossil fuels. Ultimately, the great hope of the technological optimists—solar energy—is expected to take over. One can only hope that their long-run forecasts will be better than their short-run projections have been to date. Even the most optimistic advocates of nuclear and solar energy do not suggest that these will be major sources of energy before 1990. And if the trends of the past five years continue for only another decade, major regional income shifts will have taken place.

Table 2-1.   Employment in Primary Energy Production, Selected States, 1973[a]

(1) State	(2) Total Non-agricultural Employment (000)	(3) Nonagricultural Export Base (000)	Employment (4) Coal SIC 11 & 12	(5) Oil and Gas SIC 13	(6) Total Energy (3) and (4)
1 Wyoming . . . . (32)	126 (19)	19 (16)	400 (12)	6,240 (7)	6,640 (12)
2 West Virginia . . (45)	542 (13)	173 (10)	45,015 (1)	2,811 (11)	47,826 (2)
3 Louisiana . . . (47)	1,144 (7)	227 (8)	0	41,190 (2)	41,190 (3)
4 New Mexico[b] . . (49)	345 (15)	41 (14)	850 (9)	6,885 (5)	7,735 (10)
5 Alaska[b] . . . . (3)	109 (20)	8 (19)	100 (16)	1,183 (17)	1,283 (19)
6 North Dakota . . (7)	183 (18)	12 (18)	306 (13)	1,358 (15)	1,664 (17)
7 Oklahoma[b] . . (38)	830 (10)	173 (10)	500 (11)	18,620 (3)	19,120 (6)
8 Texas . . . . . (35)	4,056 (4)	890 (4)	845 (10)	77,342 (1)	78,187 (1)
9 Kentucky . . . (43)	1,037 (8)	296 (7)	22,385 (3)	2,319 (13)	24,704 (5)
10 Utah . . . . . (42)	417 (14)	70 (13)	1,564 (7)	2,096 (16)	3,660 (16)
11 Montana . . . (33)	226 (17)	28 (15)	111 (15)	1,273 (16)	1,384 (18)
12 Kansas . . . . (15)	753 (11)	162 (11)	300 (14)	6,740 (6)	7,040 (11)
13 Colorado . . . (23)	898 (9)	153 (12)	966 (8)	5,054 (8)	6,020 (13)
14 Virginia[b] . . . (26)	1,629 (5)	408 (5)	12,640 (4)	100 (20)	12,740 (9)
15 Pennsylvania . . (21)	4,453 (2)	1,498 (2)	28,918 (2)	2,801 (12)	31,719 (4)
16 Mississippi . . . (51)	652 (12)	213 (9)	0	4,348 (9)	4,348 (15)
17 Alabama . . . (50)	1,096 (6)	340 (6)	4,849 (6)	455 (18)	5,304 (14)
18 Nevada . . . . (9)	243 (16)	14 (17)		187 (19)	187 (20)
19 California . . . (12)	7,635 (1)	1,631 (1)	0	16,721 (4)	16,721 (7)
20 Illinois . . . . (6)	4,354 (3)	1,380 (3)	10,355 (5)	3,653 (10)	14,008 (8)

Table 2–1. continued

(1) State	Energy Employment			
	(7) Coal Emp. As % Total of Energy Emp.	(8) Gas & Oil Emp. As % Total of Energy Emp.	(9) As % of Total	(10) As % of Export Base
1 Wyoming	6.02 (13)	93.98 (5)	5.27 (2)	35.00 (1)
2 West Virginia	94.12 (2)	5.87 (16)	8.82 (1)	27.50 (2)
3 Louisiana	0.00	100.00 (1)	3.60 (3)	21.80 (3)
4 New Mexico	10.99 (10)	89.01 (8)	2.24 (6)	18.80 (4)
5 Alaska	7.79 (12)	92.20 (6)	1.17 (8)	14.20 (5)
6 North Dakota	18.39 (8)	81.61 (10)	0.91 (10)	13.60 (6)
7 Oklahoma	2.61 (15)	97.38 (3)	2.30 (4)	11.00 (7)
8 Texas	1.08 (16)	98.92 (2)	1.93 (7)	8.70 (8)
9 Kentucky	90.61 (5)	9.39 (13)	2.38 (5)	8.30 (9)
10 Utah	42.73 (7)	57.28 (11)	0.88 (11)	5.20 (10)
11 Montana	8.72 (11)	91.98 (7)	0.61 (16)	4.90 (11)
12 Kansas	4.26 (14)	95.74 (4)	0.93 (9)	4.30 (12)
13 Colorado	16.05 (9)	83.95 (9)	0.67 (14)	3.90 (13)
14 Virginia	99.21 (1)	0.78 (17)	0.78 (12)	3.10 (14)
15 Pennsylvania	91.17 (4)	8.83 (14)	0.71 (13)	2.10 (15)
16 Mississippi	0.00	100.00 (1)	0.66 (15)	2.00 (16)
17 Alabama	91.42 (3)	8.58 (15)	0.48 (17)	1.50 (17)
18 Nevada	0.00	100.00 (1)	0.08 (20)	1.30 (18)
19 California	0.00	100.00 (1)	0.22 (19)	1.00 (19)
20 Illinois	73.92 (6)	26.08 (12)	0.32 (18)	1.00 (19)

Sources: Department of Commerce, *County Business Patterns*, p. 88; *Manpower Report of the President* (April 1974), p. 325.

[a]Numbers in parentheses indicate rank of per capita income in column (1) and rank of employment elsewhere.

[b]*County Business Patterns* could not disclose data for these states for columns (3) and/or (4). The figures presented were estimated by one of two techniques, either as residual subcategories of the larger mining industry or by apportioning a share of national employment on the basis of the states' share of production. The second estimating technique was only used in column (3) and was adjusted to reflect surface versus deep mining productivity differences.

A second way in which real income might be shifted from energy-consuming to energy-producing states would be the result of shifts in the location of industry induced by relative price changes. Electric energy has been a bargain not only to final consumers but to intermediate consumers who use electricity as one of their production inputs. As noted in the introductory section, only a few "power-hog" industries have been attracted to cheap energy sites in the past. But as the cost of energy goes up it will become an increasing proportion of the total cost of producing and distributing many goods. It could thus become a significant spatially-variable cost for many industries. In the past, firms in certain highly competitive industries have sought low labor-cost locations. Other industries may similarly seek low power-cost locations in the future. There could be an industrial migration similar to the North-to-South shift of the textile and hosiery industries during the late 1940s and the 1950s.

Rising energy prices will affect not only the cost of energy per se, but also the cost of transporting or transmitting energy. Increasingly, in an era of high energy prices, the nation will be forced to think of conservation, and conservation will focus attention on *net* energy consumption. Rather than using scarce and expensive energy to haul coal, for example, to manufacturing centers in the industrial Northeast, it might become economical (in terms of energy) to relocate some kinds of factories to the EP states. This could be particularly true if technological developments such as low-Btu gasification processes and gas turbines can be perfected in a reasonably short time.

It is also not unrealistic to speculate that changing energy prices will have an impact on cities. As Boulding has said:

> If indeed the automobile is replaced by public transportation, this will turn our cities outside in as the automobile turned them inside out, and we will return to the ecological patterns of the cities of 1800.[34]

But changing energy prices could reduce some of the positive externalities now enjoyed by plants located in large industrial agglomerations.[35] This could force producers to look for more advantageous locations. If enough of them were to do so, new industrial agglomerations would appear in the EP states.

CHANGING ENERGY PRICES AND REGIONAL DEVELOPMENT POLICY

We come, finally, to the central question of this paper. How will changing energy prices affect the programs of the Appalachian Regional Commission, the Economic Development Administration, and other agencies of the federal government concerned in one way or another with regional development policy? It would be wildly presumptuous to suggest that this question can be answered here. But some of the issues raised in this paper can be examined from a policy point of view. It is clearly not too early for those involved in the implementation of regional development programs to speculate about the likelihood of regional shifts in economic well-being, and the problems that these shifts will create.

Consider, first, the case of Appalachia. The problems which led to establishment of the Appalachian Regional Commission were largely limited to rural areas. Initially, the Commission poured most of its resources into the Development Highway System to provide better links between this region and the rest of the economy. One hope was that the highway system would reinforce the tendency for economic activity to cluster at "growth centers." To some extent this hope has been realized, and while the highway system is far from finished, it already has observable economic impacts on some areas. If one of the hypotheses advanced earlier in this paper should be realized— namely, that industry will be attracted to sources of relatively low-cost power—the Development Highways might prove to be far more important to this region's future economic growth than even the most optimistic highway planner envisioned when the system was first designed.

The focus of the Appalachian Regional Commission has shifted in recent years from highways to investment in human capital, that is, investment in health and educational facilities. Even a highly prosperous Appalachia will need more of this type of investment, at least for the remainder of this century.

The mandate of the Economic Development Administration is both broader and somewhat less clearly defined than that of the Appalachian Regional Commission. But the general objective has been the same—to stimulate economic development in lagging economic areas. While regional economic development programs have been in effect in the United States for almost fifteen years, and much longer in Great Britain and some of the countries of Western Europe, there is not yet complete agreement about the best approach. Economists with a strong belief in the efficacy of the market mechanism have tended to favor policies that would encourage outmigration of displaced workers from depressed areas.[36] Other economists have argued that instead of asking workers to move to jobs, jobs should be brought to structurally unemployed workers. Those who support this view urge investment in depressed areas.[37] But there is a further difference of opinion among those who advocate the second approach to the depressed-area problem. Some propose investment in directly productive activities (DPA), while others recommend investment in social overhead (SOC). Investment in DPA involves new plants, mills, mines, or other activities that would directly employ the unemployed, and indirectly stimulate employment in supporting sectors. Investment in social overhead capital involves public works: access roads, sewer systems, water-treatment facilities, or other expenditures of a type which would make depressed areas more attractive locations for private establishments. The SOC approach relies more heavily on the market mechanism than the DPA approach, since its objective is to stimulate new private investment.

But how would these approaches apply if, for example, New England becomes the "Appalachia of the future"?[38] If a substantial part of the New England manufacturing industry is placed at a competitive disadvantage because of high energy costs, many parts of the region could become depressed areas. This would not be the first spell of economic adversity for New England. The southward migration of the textile industry created severe economic distress in a substantial number of New England communities both before and after World War II.[39] But New England's earlier distress occurred at a time of growth and change in the United States economy. New industries were emerging, and one of these—electronics—was growing rapidly in New England. Expansion of the new industry helped cushion the decline of the old, although the extent to which displaced workers gravitated from the declining to the expanding industry was greatly exaggerated.[40] But there is nothing on the horizon to suggest that jobs lost in New England due to high energy prices could be replaced by some new growth sector. High energy costs will have a retarding effect on economic growth *in general*, with differential regional impacts.

What type of regional development policy would be appropriate to a region such as New England if chronic unemployment induced by high energy costs becomes a major problem? Economists with a neoclassical orientation would no doubt suggest outmigration for the displaced workers. This would assume, of course, that alternative job opportunities were available elsewhere in the nation. Other economists with a development orientation might suggest the need for federal investment in the depressed region.

But what form would this investment take? New England is already highly developed in terms of social overhead capital. Because of its location and its near-total dependence on other regions for energy, a development agency would be hard pressed to devise a plan for the long-term revival of this region under the conditions discussed above. New England is a major exporter of business and educational services, and presumably these would not be adversely affected by changes in relative energy prices. But it also is a major manufacturing region, and the entire New England manufacturing sector could be adversely affected by the kinds of relative price changes that are likely to occur. It might be necessary for New Englanders to settle for a new and lower equilibrium level of manufacturing employment.

Other areas, such as Detroit, could pose even more serious problems than New England. Detroit is a major heavy manufacturing area whose basic industry is heavily dependent upon relatively low-cost energy. Few if any industries have had as much of an impact on the American economy as the automobile industry. This is not only because the industry is linked to steel, coal, chemicals, textiles, rubber, and a vast array of others on the input side, but also because it prompted the largest single public works program ever undertaken in this country—the Interstate Highway System. If energy prices double or triple every five to ten years, Detroit will feel the impact more than any other single community. The auto industry would never have reached its present proportions without cheap energy, and expensive energy will no doubt force it to contract to a much lower equilibrium level.

Detroit's unemployment problem could be mitigated by outmigration. Indeed, as noted earlier in this paper, a limited amount of such "return" migration has already started.[41] But Detroit is likely to have a serious unemployment problem for a long time to come. What can existing regional development agencies do about situations of this kind?

Neither the Appalachian Regional Commission nor the Economic Development Administration was established with urban renewal or urban redevelopment in mind. The typical regional development approaches of investment in directly productive activities or investment in public works to stimulate growth of the private sector make little sense when applied to a city such as Detroit.

The problems of American cities do not stem entirely from the current recession or recent increases in energy prices. Their problems go far beyond the scope of this paper. Detroit has been singled out for mention because more than other cities its problems can be attributed to the changing energy price structure. Perhaps the only "solution" to America's urban problems is to admit that they are overgrown and underfinanced, and to prepare to make the necessary long-run adjustments that will be required to alter these conditions.

The most serious adjustment problem facing the nation today is one of attitudes. The American economy is faced with two problems. First, there is the problem of a highly selective recession which has hit some industries and regions much harder than others. The second problem reflects a *structural* change—the end of the era of cheap energy. It is reasonable to expect that the economy will pull out of the recession, although it may be a longer and harder pull than any we have witnessed in the past. The energy problem will not disappear. It will be with us from here on. Those political leaders who hope that Disney's First Law will take care of the energy problem are not only deluding themselves and their constituents, they are ensuring a chaotic rather than an orderly response to an entirely new kind of post-industrial world.

NOTES TO CHAPTER 2

1. For a discussion of the background and some of the consequences of this agreement, see John P. David, "Earnings, Health, Safety, and Welfare of Bituminous Coal Miners Since the Encouragement of Mechanization by the United Mine Workers of America" (Ph.D. dissertation, West Virginia University, 1972).

2. All data on production and prices used here and subsequently are from various issues of the *Minerals Yearbook*, United States Bureau of Mines.

3. Throughout this paper there are frequent references to "cheap" energy, and to "low" or "high" energy prices. The reader is asked to remember that all of these references are to *relative* prices.

4. John M. Goshko, "OPEC—From Ineptitude to World's Most Powerful Cartel," *Washington Post*, December 22, 1974.

5. Ibid.

6. Richard B. Mancke, *The Failure of U.S. Energy Policy* (New York: Columbia University Press, 1974), p. 28.

7. M.A. Adelman, "Is the Oil Shortage Real?" *Foreign Policy*, Winter 1972/73, p. 70.

8. Ibid.

9. Ibid.

10. Ibid.

11. Louis Kraar, "OPEC Is Starting to Feel the Pressure," *Fortune*, May 1975, pp. 186ff.

12. See, for example, a statement by John Hill, Deputy Administrator of the Federal Energy Administration, *Wall Street Journal*, May 20, 1975.

13. Anthony Parisi, "The Pressure on OPEC to Raise Oil Prices Again," *Business Week*, May 26, 1975, p. 28. See also Juan deOnis, "OPEC, Facing Real Losses, May Impose Oil Increases," *Wall Street Journal*, June 1, 1975.

14. Ibid.

15. Georgescu's latest publication, which summarizes much of his recent work, is "Energy and Economic Myths," *Southern Economic Journal*, January 1975, pp. 347–81. The formal analytical basis of Georgescu's ideas are given in *The Entropy Law and The Economic Process* (Cambridge, Mass.: Harvard Uni-

versity Press, 1971). A highly readable condensation of his basic views is given in "The Entropy Law and The Economic Problem," Distinguished Lecture Series No. 1, Department of Economics, University of Alabama, 1971, reprinted in *The Ecologist*, July 1972, pp. 13—18.

16. Georgescu-Roegen, "Energy and Economic Myths," p. 348.

17. Ibid.

18. Ibid., p. 349.

19. Ibid.

20. Ibid., p. 352.

21. Ibid., p. 378.

22. For a concise summary of the results of six studies of potential oil and gas exhaustion, see Deborah Shapley, "Senate Studies Predict U.S. Oil 'Exhaustion,' " *Science*, March 21, 1975, p. 1064.

23. "Oil and Gas Resources: Academy Calls USGS Math 'Misleading,' " *Science*, February 28, 1975, p. 725.

24. See Donella H. Meadows et al., *The Limits to Growth* (New York: Universe Books, 1972). For a technical discussion of the forecasting method used, see Jay Forrester, *World Dynamics* (Cambridge, Mass.: Wright-Allen, 1971).

25. See Karl Kaysen, "The Computer That Printed Out W*O*L*F*," *Foreign Affairs*, July 1972, pp. 660—68; and Robert M. Solow, "The Economics of Resources and the Resources of Economics" (Richard T. Ely lecture), *American Economic Review*, May 1974, p. 1.

26. Georgescu-Roegen, "Energy and Economic Myths," p. 359.

27. Mancke, op. cit., p. 131.

28. See end note 31, below.

29. For a discussion of this technique and its prospects, see William H. Miernyk and John T. Sears, *Air Pollution Abatement and Regional Economic Development* (Lexington, Mass.: D.C. Heath and Company, 1974).

30. For a brief report which summarizes the early ERDA philosophy of energy research, see *An Analysis Identifying Issues in the Fiscal Year 1976 ERDA Budget*, Report prepared by the Office of Technology Assessment for the Committee on Science and Technology, U.S. House of Representatives; Committee on Interior and Insular Affairs, U.S. Senate; and the Joint Committee on Atomic Energy, 94th Congress, 1st session (Washington, D.C.: Government Printing Office, 1975).

31. For example: Chauncey Starr, "Energy and Power," *Scientific American*, September 1971, pp. 37—49; Edward A. Hudson and Dale W. Jorgenson, "U.S. Energy Policy and Economic Growth," *Bell Journal of Economics and Management Science*, Autumn 1974, pp. 461—514; *Exploring Energy Choices: A Preliminary Report*, Energy Policy Project of the Ford Foundation (1974), especially Section 7, pp. 39—53. Projections with some regional details are given in the *Project Independence Report* issued by the Federal Energy Administration in 1974.

32. The expansion of surface mining raises ecological issues which are important, but which are extraneous to, the present discussion.

33. For additional details, see William H. Miernyk, "Regional Employment Impacts of Rising Energy Prices," *Labor Law Journal*, August 1975, pp. 518—23.

34. Kenneth E. Boulding, "The Social System in the Energy Crisis," *Science*, April 19, 1974, p. 257.

35. On the other hand, if public transportation is widely accepted, changing energy prices could have the opposite effect. The question remains: What kinds of economic activities would be attracted to EC cities if they are at a serious comparative disadvantage in terms of energy cost?

36. See, for example, Niles Hansen, *Rural Poverty and the Urban Crisis* (Bloomington: Indiana University Press, 1970), especially pp. 271–87.

37. For an extensive discussion of the central issues involved in this debate, see Marina V.N. Whitman, "Place Prosperity and People Prosperity: The Delineation of Optimum Policy Areas," in Perlman, Leven, and Chinitz, eds., *Spatial, Regional and Population Economics* (New York: Gordon and Breach, 1972), pp. 359–93.

38. This possibility was suggested by Leonard J. Hausman of Brandeis University at the session on "Economic Development and Manpower Problems" of the Industrial Relations Research Association Annual Spring Meeting, Hartford, Connecticut, May 9, 1975.

39. For a general discussion, see Arthur A. Bright, Jr. and George Ellis, eds., *The Economic State of New England* (New Haven: Yale University Press, 1954). See also William H. Miernyk, *Depressed Industrial Areas—A National Problem* (Washington, D.C.: National Planning Association, Planning Pamphlet No. 98, 1957).

40. On this, see William H. Miernyk, *Inter-Industry Labor Mobility* (Boston: Northeastern University, Bureau of Business and Economic Research, 1955).

41. See "Coal Rush Lures 'Hillbillies' Home From Northern Jobs," *New York Times*, May 4, 1975.

✳ *Chapter 3*

The Northeast Isn't What It Used to Be

William H. Miernyk

The most striking feature in the history of American manufacturing is the enduring strength of the Northeast . . . even today (1960) the great industrial belt in the Northeast continues to dominate the regional structure of American manufacturers much as it did at the beginning of this century—
Perloff et al., *Regions, Resources, and Economic Growth.*

AN HISTORICAL SKETCH OF THE NORTHEAST ECONOMY

Historically, the Northeast has been an important part of the manufacturing base of a growing U.S. economy.[1] As a region, the Northeast has been an importer of raw materials and energy, and an exporter of finished products. The quotation from Perloff and his associates at the beginning of this paper is taken from a comprehensive regional economic analysis of the American economy conducted during the late 1950's [26]. It was primarily concerned with the *changing* regional structure of the American economy. But despite the changes which these authors observed and documented—including regional shifts in various manufacturing activities—the Northeast continued to be regarded by them as the nation's "manufacturing belt."

In the literature of development economics—whether regional or national—manufacturing has always been considered to be the most important component of an economy's export base. This is because

This chapter is excerpted from a longer paper published in *Balanced Growth for the Northeast*, Proceedings of a Conference of Legislative Leaders on the Future of the Northeast (Albany, New York: New York State Senate, 1975).

of the high income and employment multipliers associated with manufacturing. These multipliers, in turn, are related to the high degree of interdependence between a manufacturing firm and other business establishments in the region in which it is located. It is intuitively clear that any change in that firm's fortunes will have widespread repercussions throughout the regional economy.

The economy of the Northeast has not rested entirely on its manufacturing base. Quite early New York became the financial center of the nation; a number of major insurance companies in the Northeast have long served national markets. Also because of the concentration of prestigious colleges and universities in the Northeast, the region has long been an exporter of educational services.

One cannot say that the Northeast is now favorably located with respect to national markets. But the Northeast was the cradle of American manufacturing industry, and for many years it benefited from the inertia of the initial location of industry in the region. Not all of the states of the Northeast shared equally in this prosperity; the three northern New England states, for example, consistently have had per capita incomes below the national average. But New York and Connecticut have regularly been at or close to the top in per capita personal income among the 48 conterminous states, while Massachusetts and Pennsylvania have also ranked well above the national average. Until recently the "terms of trade" between the Northeast and the nation as a whole were favorable to this region.[2] The consistently high per capita income of the northeastern states strongly suggests that this region regularly enjoyed a "favorable" balance of trade with the rest of the nation.

THE IMPACT OF RECENT ECONOMIC CHANGES ON THE NORTHEAST

The Northeast is no stranger to economic adversity. This is particularly true of New England [4, 13, 19]. The textile exodus which started between the two World Wars, and which was resumed at the end of World War II, had a major impact on some areas in the Northeast. There was widespread and persistent unemployment in many of the textile manufacturing centers of the Northeast following the recession of 1948–1949 during which many older mills were liquidated while new mills were being opened in the South. But the region has always bounced back. New activities have been substituted for old. In New England, for example, the growth and rapid expansion of the electronics industry took place at a time when regional textile employment and production was declining.[3]

Another kind of employment substitution that occurred during this period was represented by the shift from manufacturing to financial and other services. This phenomenon led to a vigorous debate among a number of economists in New England.[4] Some, such as the Alfred C. Neal—a former vice president of the Federal Reserve Bank of Boston—claimed that the shift of employment dependence from manufacturing and other "primary" activities to the services and other "tertiary" activities was the manifestation of healthy economic growth. Others, notably Seymour Harris of Harvard, argued that the loss of manufacturing jobs was a sign of economic weakness, and that the offsetting growth of service jobs would not be sufficient to compensate for the region's loss of its comparative advantages in manufacturing [13, pp. 279–87]. But whatever the merits of the arguments on either side in this debate—and, as is true in many debates of this kind, one can see retrospectively that there was some truth on both sides—the Northeast always seemed to rebound from periods of economic distress.

Since the early 1970s—and particularly since late 1973—the national economy has been in a recession. This recession has differed both in kind and degree from earlier "business cycles." It has been a highly selective recession. Some sectors of the economy—the automobile industry is a notable example—have experienced severe cutbacks in production and employment. Other sectors have continued to grow, although obviously at slower rates than they would have grown in the absence of a recession. The recession has also been selective on a geographic basis. Some major urban areas—notably New York City and Detroit—are in the throes of severe financial crises. And the entire Northeast (with the possible exception of Pennsylvania) has been on the borderline of financial crisis. Meanwhile, other regions have experienced declining unemployment rates and fairly rapid increases in per capita personal income.

The explanation of the selective effects of the present recession is that two sets of economic forces are at work on the American economy. One set is the conventional cyclical forces that have caused recessions in the past. But at the same time, since October 1973, the American economy has felt the impact of a *structural* change in world export prices.[5] This structural change in export prices was, of course, induced by actions of the Organization of Petroleum Exporting Countries (OPEC) when oil prices quadrupled following the 1973 embargo.

The impact of rising energy costs has been particularly severe on the Northeast. This is partly because of the heavy dependence of parts of this region on imported residual fuel oil [see 27, p. 18]. But it is also partly because the Northeast pays more for domestically produced energy due to the added cost of transporting energy to the region. This is illustrated in Table 3–1, which shows the average cost of domestic and imported natural gas at the point of consumption. The data in Table 3–1 are for 1972, well before the effects of the oil embargo had been felt by the American economy. At that time, however, the average cost of natural gas in New England was over two and one-half times the national average, while in the Middle Atlantic states it was almost 180 percent of this average. In the long run there tends to be a rough "parity" in energy prices. So what has been said about natural gas prices in the Northeast and in other regions applies in general to other forms of energy.

The future of the Northeast economy will depend to a large extent on the relative prices of its imports and exports; that is, its terms of trade with the rest of the domestic economy (as well as the rest of the world). And the interregional terms of trade between the Northeast and other regions will likely be determined to a large extent by future energy prices.

In the past, the Northeast has always rebounded from economic adversity. But will it be able to make the necessary adjustments this time? If indeed we have come to the end of the era of cheap energy in the United States, the industrial preeminence of the Northeast could be challenged. In this case the regional economy might not fully recover from the present recession; it might be forced to settle at a lower equilibrium level. This, in fact, is what has happened to

Table 3–1.   Average Value of Natural Gas, at Point of Consumption, by Region, 1963–1972[a] *(cents per thousand cubic feet)*

Region	1963	1972	Percent of 1972 U.S. Average	Change Cents	Change Percent
New England	151.3¢	183.2¢	268.6%	31.9	21.08
Middle Atlantic	100.4	121.2	177.7	20.8	20.72
East North Central	75.5	85.5	125.4	10.0	13.25
West North Central	48.6	65.5	96.0	16.9	34.77
South Atlantic	71.8	83.0	121.7	11.2	15.60
East South Central	48.3	61.0	89.4	12.7	26.29
West South Central	23.2	34.7	50.9	11.5	49.57
Mountain	37.1	56.7	83.1	19.6	52.83
Pacific	57.8	69.6	102.1	11.8	20.42
U.S. Total	51.2	68.2	100.0	17.0	33.20

Source: *Minerals Year Book*, 1963 and 1972.

[a] Average of foreign and domestic natural gas.

some of the former textile centers of the region. In the course of a generation or more, their status as "depressed" areas has changed. But this was partly the result of attrition. While some of the former textile centers no longer have excessively high unemployment rates, they are smaller than they were at the peak of their development.

Ultimately, what happens in the Northeast will depend on *worldwide* relationships between people and resources. The following section might appear to be a lengthy digression, but it is not. Although the issues discussed have implications for all regions, and for all nations, they bear directly on the matters to be discussed in later sections of this paper.

POPULATION, NATURAL RESOURCES AND ENERGY: A BRIEF SUMMARY OF THREE VIEWS

In a number of ways the Northeast resembles the insular economy of Great Britain. And with appropriate lag adjustments its economic history parallels that of Great Britain following the Industrial Revolution. The first view to be discussed in this section had its origin in the early days of the Victorian period, which gave birth to "the idea of progress" [5, pp. 34–52].[6]

Neo-Victorians

The fundamentals of the idea of progress go back to the early seventeenth-century work of Descartes [29, p. 382]. But explicit statements of the idea of progress—of unlimited improvement in mankind's material well-being due to advances in science and technology—came not only from scientists but from the pens of Victorian authors and poets such as Macaulay and Tennyson. Thus the rubric *neo-Victorian* for those modern futurists who insist that there will be no end to progress—that mankind's earthly destiny is to enjoy steady improvement in productive capabilities with continuously rising standards of living.

The neo-Victorians are, essentially, technological optimists. In their view there is no problem—economic or otherwise—that cannot be solved by science and technology. One view of the energy crisis leads to the conclusion that we really have nothing to worry about. Present economic problems which were engendered by the quadrupling of world petroleum prices in 1973 are transitory. This is because all cartels contain the seeds of their own destruction; and OPEC, some neo-Victorians say, will be no exception.

The view that OPEC will collapse, and thus bring down the structure of high energy prices, is not, of course, based on a technological argument. Indeed, this view has been advanced by economists rather than scientists or engineers. The belief that cartels cannot survive is based on an implicit faith that competition will conquer all. The economists who support this view believe that competitive markets are widespread and effective. They believe that in the long run the forces of competition cannot be frustrated.[7] But these economists also appear to believe that the supply of energy resources is virtually unlimited. If one source of energy begins to peter out, scientists and engineers will develop alternative or substitute sources.

The strictly technological argument that there is no need to worry about high energy prices takes a somewhat different tack from the economic view given above. Basically it is as follows: OPEC has forced energy prices up. But present high fossil fuel prices will hasten the day of the "breeder" reactor and cheap nuclear energy. Later, we will have nuclear fusion. And finally the *deus ex machina*, solar energy, will appear. Then far from having an energy shortage we will have virtually unlimited and thus very cheap (if not entirely free) energy.

Stripped to its essentials, the neo-Victorian view is theological rather than analytical or empirical. True, it is a secular theology, but the idea of unlimited material progress for mankind is as much an article of faith as are the conventional religious concepts of immortality and salvation. The basis of the faith in the technological solution to contemporary problems appears to be nothing more than a willingness to extrapolate from the past. Science and technology have solved the problems of the past, and will solve the problems of the future.

Neo-Victorians are not likely to be impressed by the arguments of others. Their behavior is similar to that of a Victorian character created by Charles Dickens:

> Thus happily acquainted with his own merit and importance, Mr. Podsnap settled that whatever he put behind him he put out of existence. There was a dignified conclusiveness—not to add a grand convenience—in his way of getting rid of disagreeables which had done much toward establishing Mr. Podsnap in his lofty place in Mr. Podsnap's satisfaction. "I don't want to know about it; I don't choose to discuss it; I don't admit it!" Mr. Podsnap had even acquired a peculiar flourish of his right arm in often clearing the world of its most difficult problems, by sweeping them behind him (and consequently sheer away) with those words and a flushed face. For they affronted him [7, p. 143].

This indomitable attitude toward the correctness of Mr. Podsnap's views was labeled "Podsnappery" by Mr. Dickens. Many neo-Victorians appear to be equally affronted by the suggestion that today's problems are anything but transitory. Podsnappery is far from dead. But what if the neo-Victorians should turn out to be right? Then the Northeast has nothing to worry about—*in the long run.*

THE NEO-MALTHUSIAN VIEW

Thomas Robert Malthus was the English cleric and economist who believed that mankind was destined to engage in a constant race with starvation. In 1798, Malthus published a brief tract in which he argued that population would constantly tend to outrun the means of subsistence. Population would grow geometrically, he argued, and the means of subsistence would follow an arithmetic progression. The pamphlet was enlarged and revised and published in 1803 as the *Essay Population*, a book that was to make the name of Malthus a household word.

What Malthus failed to take into account in his simple model was technology. He did not foresee the quantum increases in production which permitted a steadily declining proportion of the population of industrialized nations to feed steadily increasing populations—and to produce vast surpluses that could be sent to other less-developed nations. There increases in production were the result of massive gains in productivity which were, in turn, made possible by technological progress.

Technological optimists point to evidence of this kind in support of their view that there is no ultimate ceiling on productivity. But there are scholars who argue that because of man's fecundity Malthus will ultimately prove to be right. Interest in the Malthusian argument has waxed and waned over the past century and a half. The most recent revival of neo-Malthusianism followed publication of *The Limits to Growth* in 1972 [17]. The study reported in this book was sponsored by an organization, known as the Club of Rome, concerned with various problems facing man in a rapidly changing world.

The neo-Malthusian view is that the world is rapidly using up its available resources in relation to population. Sometime during the twenty-first century there will be an "overpopulation crisis." The authors of *The Limits to Growth*, Dennis and Donella Meadows, foresee a continued growth of population until about the year 2100. Available resources will permit growth to proceed until then. But at some point toward the end of the next century there simply will not be enough resources to support the number of persons who will be alive at that time. Following the crisis there will be a sharp population decline as the world adjusts to its remaining stock of resources.

The gloomy prophecies reported in *The Limits to Growth* were based on a highly sophisticated computer simulation model—known as World 3—using a simulation technique originally developed by Jay Forrester of MIT [18]. There has been some misunderstanding about the nature of this model. *The Limits to Growth* was bitterly criticized by mainstream economists who were convinced that despite the sophistication of the model its conclusions were no more tenable than those reached by Malthus almost two centuries ago. As the more detailed and more technical treatment of the model points out, however, the Meadows team is more concerned with the structure of the model than with the specific forecasts it has made. They are willing to admit that their projections could be off in timing, but they believe that "a system that possesses . . . the three characteristics of: rapid growth, environmental limits, and feedback delays—is inherently unstable" [18, p. 562]. While it might not be possible to pinpoint the coming population crisis, they continue, we are not likely to witness changes in lifestyle—or even in technological and social policies—of the direction and magnitude that would be needed to postpone the crisis indefinitely.

Despite the severe criticism of the first book published by the Meadows team, the ideas behind *The Limits to Growth* die hard. The first of a series of biennial conferences designed to further explore these ideas was held at the University of Houston on October 19–21, 1975. Participants included supporters of the conclusions reached by the Meadows team, as well as some of the nation's better-known technological optimists.

If the neo-Malthusians are right, the outlook for the Northeast is grim. But then the outlook for the entire world is grim, and the Northeast has one consolation: it has a lot of accumulated wealth on which it can draw between now and the time of the ultimate collapse.

THE MODERATE PESSIMISM OF GEORGESCU-ROEGEN

Nicholas Georgescu-Roegen has been hailed as an "economist's economist" by no less a luminary than Nobel Laureate Paul Samuelson. But Georgescu-Roegen's recent work—which may be the most important work going on in modern economics—has been largely ignored by most conventional economists who are, among other things, technological optimists. Those who do not believe that OPEC will collapse, and that cheap energy will once again be restored to the world, have a deep and abiding faith that technology will solve the world's energy problems.

It is important to stress that Georgescu is not prophesying impending disaster. The policy suggestions that emerge from his hard-hitting analysis of the real world deal with the *conservation* of energy and resources in a realistic manner. By way of contrast, some economists who dwell in the imaginary world described by the "perfectly competitive model" believe that unless a resource is sold at the "competitive price" it is possible to *overconserve* the resource.[9] A neo-Victorian economist might regard recycling, for example, as nothing more than a palliative. But a number of geologists and earth scientists have argued that recycling is the only way that the present stock of mineral resources can be extended over time.

The significance of Georgescu's work for the present paper is that it provides a solid economic and physical basis for the conclusion that the world has reached the end of the era of cheap energy. The structural change in world export prices that occurred late in 1973 was not a transitory phenomenon. From here on the world can expect energy prices to climb. More significantly, in the short run energy prices are likely to increase more than other prices; that is, the *structure* of prices will continue to change. These assumptions are basic to the following analysis of the impact of changing energy prices on the economic future of the Northeast.

THE CONSEQUENCES OF HIGH ENERGY PRICES FOR THE NORTHEAST

The Northeast, with the exception of Pennsylvania, is "energy poor." It has to import all but a small fraction of the energy it consumes, both for final consumption and for use by the region's reproduction processes.

For the next 15 to 20 years the United States will rely increasingly on fossil fuels. OPEC is not on the verge of collapse, and crude oil prices will rise further. Because demand will increase faster than supply, coal and natural gas prices also will rise. Also, regardless of the method of generation, the cost of electric energy will go up steadily. As already noted (Table 3–1), the Northeast has been at an energy-cost disadvantage because of the cost of transporting to the region. There is nothing on the horizon to suggest that this situation is going to change. The relationship may even worsen slightly since expensive energy must now be used to transport energy to the Northeast.

The production and consumption of domestically produced basic energy by regions is given in Tables 3–2 through 3–4. Table 3–5 shows the production and consumption of electric energy by region, and Table 3–6 shows the next export position of each region for the various types of energy discussed.[10]

Table 3–2. Production and Consumption of Coal, by Region, 1963–1972 (thousands of net tons)

Region	Production 1963	Production 1972	Change Tons	Change Percent	Consumption 1963	Consumption 1972	Change Tons	Change Percent
New England	71,501	75,939	4,438	6.21	10,017	1,522	−8,495	−84.81
Middle Atlantic					79,492	78,998	−494	−.62
(MA–Pa.)	(0)	(0)	(0)	(0)	(29,291)	(14,480)	(−14,811)	(−50.57)
East North Central	103,626	142,439	38,813	37.45	164,423	206,504	42,081	25.59
West North Central	7,971	13,261	5,290	66.37	23,242	39,587	16,345	70.33
South Atlantic	164,266	159,411	−4,855	−2.96	63,816	96,907	33,091	51.85
(SA + Pa.)	(235,767)	(235,350)	(−417)	(−0.18)	(114,017)	(161,425)	(−47,408)	(41.58)
East South Central	95,830	153,262	57,432	59.93	47,418	78,843	31,425	66.27
West South Central	1,229	7,097	5,868	477.46	802	930	128	15.96
Mountain	13,462	40,675	27,213	202.15	10,823	26,330	15,507	143.28
Pacific	1,043	3,302	2,259	216.59	3,373	5,352	2,979	125.54
U.S. Total	458,928	595,386	136,458	29.73	403,406	534,973	131,567	32.61

Source: Minerals Year Book, 1963, 1972.

Table 3–3. Production and Consumption of Natural Gas, by Region, 1963–1972 *(million cubic feet)*

Region	Marketed Production				Consumption			
	1963	1972	Change Number	Change Percent	1963	1972	Change Number	Change Percent
New England					151,577	260,049	108,472	71.6
Middle Atlantic	96,619	77,637	-18,982	-19.6	1,280,617	1,843,049	562,432	43.9
(MA–Pa.)	(3,962)	(3,679)	(-283)	(-7.1)	(695,191)	(1,014,018)	(318,827)	(45.9)
East North Central	79,412	125,765	46,353	58.4	2,369,668	4,132,895	1,763,227	74.4
West North Central	778,895	925,227	146,332	18.8	1,326,945	2,088,836	761,891	57.4
South Atlantic	213,976	233,503	19,527	9.1	893,277	1,533,501	640,224	71.7
(SA + Pa.)	(306,633)	(307,461)	(828)	(0.3)	(1,478,703)	(2,362,532)	(883,829)	(59.8)
East South Central	251,708	171,306	-80,402	-31.9	778,293	1,196,300	418,007	53.7
West South Central	11,443,445	18,603,927	7,160,482	62.6	5,223,682	8,024,175	2,800,493	53.6
Mountain	1,231,624	1,781,459	549,835	44.6	927,897	1,364,726	436,829	47.1
Pacific	650,984	612,874	-38,110	-5.9	1,688,524	2,565,914	877,390	52.0
U.S. Total	14,746,663	22,531,698	7,785,035	52.8	14,640,480	23,009,445	8,368,965	57.1

Source: *Minerals Year Book*, 1963, 1972.

Table 3–4. Production and Consumption of Crude Petroleum, by Region, 1963–1972 *(thousands of barrels)*

Region	Production		Change		Consumption		Change	
	1963	*1972*	*Barrels*	*Percent*	*1963*	*1972*	*Barrels*	*Percent*
New England								
Middle Atlantic	6,892	4,459	-2,433	-35.30	7,123	4,561	-2,562	-35.97
(MA–Pa.)	(1,929)	(1,018)	(-911)	(-47.23)	(1,929)	(1,018)	(-911)	(-47.23)
East North Central	107,344	63,352	-43,992	-40.98	108,053	66,971	-41,082	-38.02
West North Central	156,080	103,352	-52,728	-33.78	156,601	105,816	-50,785	-32.43
South Atlantic	3,710	19,574	15,864	427.60	3,896	19,135	15,239	391.14
(SA + Pa.)	(8,673)	(23,015)	(14,342)	(165.36)	(9,090)	(22,678)	(13,588)	(149.48)
East South Central	86,989	80,934	-6,055	-6.96	87,320	81,469	-5,851	-6.70
West South Central	1,723,447	2,419,664	696,217	40.40	1,732,911	2,425,370	692,459	39.96
Mountain	356,788	344,118	-12,670	-3.55	358,819	348,510	-10,309	-2.87
Pacific	311,473	419,915	108,442	34.82	312,406	420,600	108,194	34.63
U.S. Total	2,752,723	3,455,368	702,645	25.53	2,767,129	3,472,432	705,303	25.49
Foreign Imports					412,904	807,324	457,420	110.78
U.S. Total & Imports					3,180,033	4,279,756	1,162,723	36.56

Source: *Minerals Year Book,* 1963, 1972.

Table 3–5. Production and Consumption of Electric Energy, by Region, 1960–1973 (millions of Kwh)

Region	Production				Consumption			
			Change				Change	
	1960	1973	Number	Percent	1960	1973	Number	Percent
New England	26,161	66,454	40,293	154.0	26,921	69,140	42,219	156.8
Middle Atlantic	102,786	226,514	123,728	120.4	103,231	235,534	132,303	128.2
(MA–Pa.)	(63,470)	(130,752)	(67,282)	(106.0)	(63,714)	(145,496)	(81,782)	(128.4)
East North Central	150,485	319,495	169,010	112.3	149,402	340,374	190,972	127.8
West North Central	39,727	107,302	67,575	170.1	42,210	108,295	66,085	156.6
South Atlantic	94,450	305,626	211,176	223.6	90,950	284,919	193,969	213.3
(SA + Pa.)	(133,766)	(401,388)	(276,622)	(200.1)	(130,467)	(374,957)	(244,490)	(187.4)
East South Central	80,092	165,585	85,493	106.7	88,018	164,370	76,352	86.7
West South Central	58,774	202,586	143,812	244.7	58,086	199,557	141,471	243.6
Mountain	33,413	89,206	55,793	167.0	30,865	78,689	47,824	154.9
Pacific	100,619	225,066	124,447	123.7	96,814	226,047	129,233	133.5
U.S. Total	686,507	1,707,834	1,021,327	148.8	686,497	1,706,925	1,020,428	148.7

Source: Federal Power Commission, *Electric Power Statistics*, 1960, 1973.

Table 3-6. Net Exports of Energy Resources, by Region

Region	Coal 1963	Coal 1972	Natural Gas 1963	Natural Gas 1972	Crude Petroleum 1963	Crude Petroleum 1972	Electric Energy 1960	Electric Energy 1973
New England	-10,017	-1,522	-151,577	-260,049	-231	-102	-760	-2,686
Middle Atlantic	-7,991	-3,059	-1,183,998	-1,765,412			-445	-9,020
(MA-Pa.)	(-29,291)	(-14,480)	(-691,229)	(-1,010,339)	(0)	(0)	(-244)	(-14,744)
East North Central	-60,797	-64,065	-2,290,256	-4,007,130	-709	-3,619	+1,083	-20,879
West North Central	-15,271	-26,326	-548,050	-1,163,609	-521	-2,464	-2,483	-993
South Atlantic	+100,450	+62,504	-679,301	-1,299,998	-186	+439	+3,500	+20,707
(SA + Pa.)	(+121,750)	(+73,925)	(-1,172,070)	(-2,055,071)	(-417)	(+337)	(-3,299)	(-26,431)
East South Central	+48,412	+74,419	-526,585	-1,024,994	-331	-535	-7,926	+1,215
West South Central	+427	+6,167	+6,219,763	+10,579,752	-9,464	-5,706	+688	+3,029
Mountain	+2,639	+14,345	+303,727	+416,733	-2,031	-4,392	+2,548	+10,517
Pacific	-2,330	-2,050	-1,037,540	-1,953,040	-933	-685	3,805	-981
U.S. Total	+55,522	+60,413			-14,406	-17,064		
Statistical Discrepancy	2,791	162	106,183	-477,747			10	909
Foreign Imports					+412,904	+807,324		
U.S. Total & Imports					-427,310	-824,388		

Sources: *Minerals Year Book*, 1963, 1972 and Federal Power Commission, *Electric Power Statistics*, 1960, 1973.

Because Pennsylvania is an important energy-producing state, the inclusion of Pennsylvania in the Middle Atlantic region could be misleading as far as the rest of the Northeast is concerned. Thus in all of the tables the Middle Atlantic region is shown both with and without Pennsylvania, and the South Atlantic region—which Pennsylvania borders—is also shown including and excluding Pennsylvania. In all cases where these adjustments are made, the numbers are given in parentheses.

Table 3—2 shows that when Pennsylvania is excluded from the Northeast, this heavily industrialized region produces no coal. It is the only region which is entirely dependent on imports for this basic source of energy. For some time the Northeast has been reducing its dependence on coal, but it has done this by increasing its dependence on imported crude and residual fuel oil [27]. At the other end of the scale, the South Atlantic, East South Central, and Mountain regions are major exporters of coal, and there were larger increases in their export between 1963 and 1972.

The Northeast's energy situation is only slightly better in the case of natural gas (Table 3—3). Again, New England produces none of the natural gas it consumes. The Middle Atlantic region is a substantial producer when Pennsylvania is included in the region, but the region's production is greatly reduced when Pennsylvania is excluded. The only regions that were major exporters of natural gas in 1972 were the West South Central—the largest by far—and the Mountain region.

Table 3—4 shows the production and consumption of *domestically produced* crude petroleum by region. New England purchased no crude petroleum from domestic producers in either 1963 or 1972. The Middle Atlantic region also consumed a relatively small amount of domestically produced petroleum. More than anything else, Table 3—4 shows the heavy reliance of the United States—and especially the Northeast—on imported petroleum. The only region to show a net export surplus in 1972 (Table 3—6) was the South Atlantic region. All other regions, including those that produced crude petroleum, had to supplement their domestic production by substantial imports.

As Table 3—5 shows, all regions produced electric energy. But some produce more than they consume. The Northeast shows a deficit. It has to import a substantial amount of electricity, and its imports increased somewhat during the brief period covered by Table 3—5. The South Atlantic region is the major exporter of electric energy, and some of this no doubt is sold to the Northeast.

These tables probably contain few surprises for anyone familiar with the economy of the Northeast. It would have been more useful to show the same information on a state-by-state basis, but the large tables required would have been cumbersome. Even the compressed tables show enough, however, to indicate what is likely to happen to the Northeast if, in fact, energy prices rise more rapidly than other prices during the next 15 to 20 years.

It will be useful to think in the terms that economists use when analyzing international trade and discussing the impacts of the changing energy price structure on the Northeast. Each region in the national economy produces a "bundle" of goods and services. Some of these are for domestic consumption; others are for export. If the bundle of goods and services exported by New England, for example, is worth more than the bundle of goods and services it must import, we say that the "terms of trade" favor New England. This region has, in other words, a "favorable balance of trade," and is earning money on its export account.[11] But if the cost of the bundle of imports which New England must purchase to produce and sell export goods and services goes up faster than the prices of its exports, the terms of trade will have turned against New England.[12]

If a large part of the increase in import prices is a result of higher energy prices, there will be a shift in *real* income from the Northeast to the nation's energy-producing states. Indeed, there is evidence that this shift has already started. First, there have been rapid gains in personal income in recent years in such energy-producing states as Alaska, New Mexico, West Virginia, Wyoming, and Kentucky. Meanwhile, Massachusetts, New York, and Connecticut ranked at the bottom of the income-change table (see *Survey of Current Business*, April 1975, p. 24). There is also more direct evidence that shifts in real income have taken place.

Windfall gains to the coal and domestic petroleum and natural gas industries have been calculated using the most recent national input-output table published by the U.S. Department of Commerce, and the latest price published by the Bureau of Labor Statistics. The results are summarized in Figures 3-1 and 3-2.

The numbers at the left-hand side of Figure 3-1 refer to the input-output classification of industries. Only those relevant to the present paper will be identified here. Sector 7 is coal mining; Sector 31 is petroleum refining and related industries; and Sector 8 is crude petroleum and natural gas. The increases in value added in these three sectors stand in marked contrast to those in the remaining processing sector of the economy.

The direct and indirect gains to petroleum and natural gas from price increases between 1967 and 1974 amounted to more than

Figure 3–1. Relative Impacts of Price Changes on Value Added
(percent change, 1967–1974)

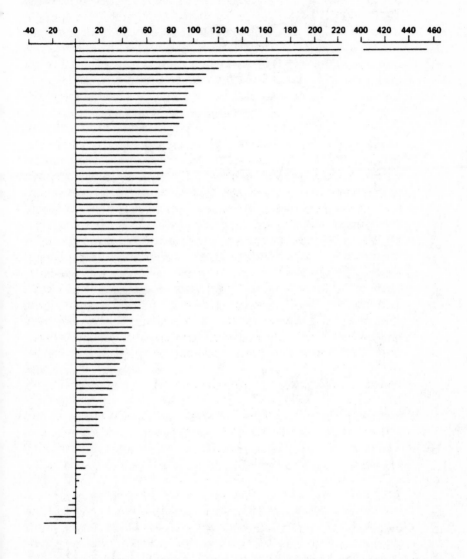

Source: Miernyk, "Some Regional Impacts of the Rising Costs of Energy," Annual Meeting, Regional Science Association, Cambridge, Massachusetts, November 7–9, 1975.

Figure 3-2.   Distribution of Incremental Value Added, by State

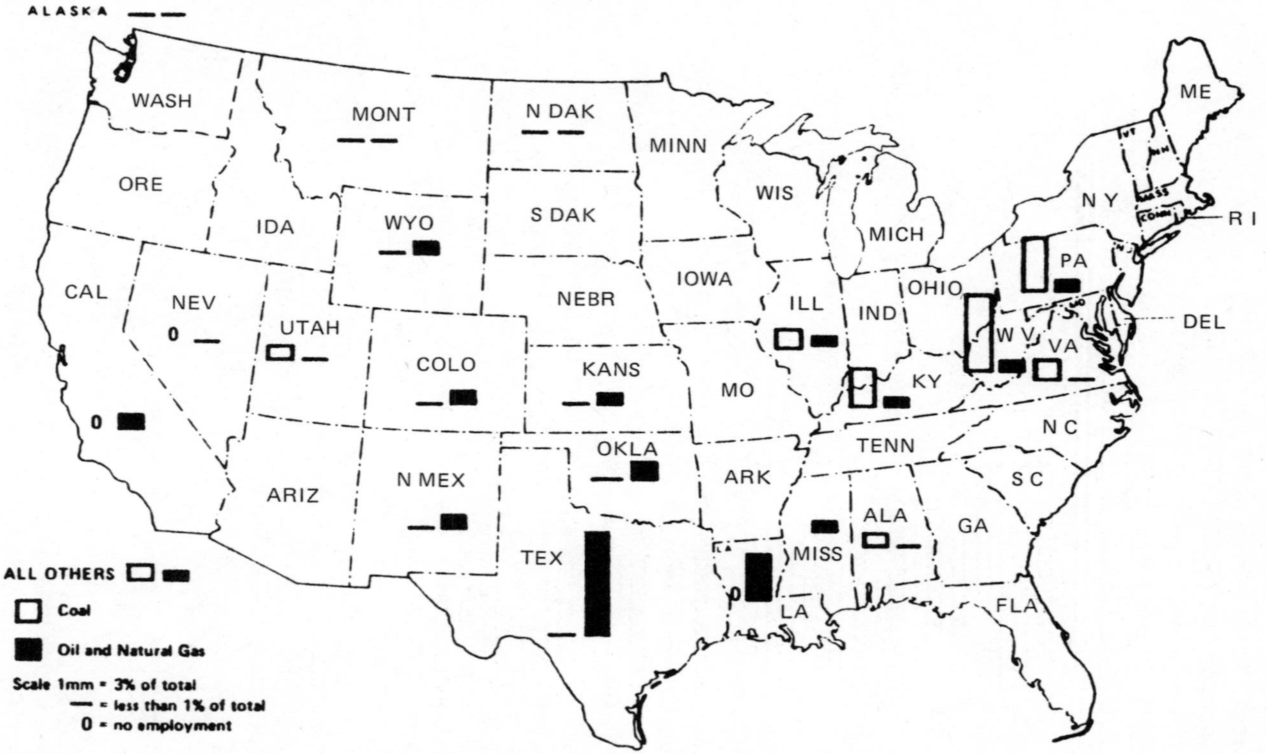

ALL OTHERS ▢   Coal
           ■   Oil and Natural Gas

Scale 1mm = 3% of total
      ▬ = less than 1% of total
      0 = no employment

Source: Miernyk, "Some Regional Impacts of the Rising Costs of Energy," Annual Meeting, Regional Science Association, Cambridge, Massachusetts, November 7–9, 1975.

$15.7 billion. The windfall gains to coal came to more than $8.5 billion during the same period. No doubt there have been further gains in both of these sectors since then.

Figure 3–2 shows how these gains were distributed on a geographic basis; the height of each bar represents the percentage share of the amounts given above. Substantial windfall returns accrued to such states as West Virginia, Kentucky, Louisiana, Oklahoma, and Texas— all states with below-average per capita income in 1974. The only state in the Northeast to participate in these windfall gains was Pennsylvania, which ranked second to West Virginia in terms of the value added to coal between 1967 and 1974. This map tells only half the story; it shows which states received the gains, but it doesn't show how the higher energy bills were divided among the remaining states. It is clear, however, that a substantial part of these bills had to be paid by the Northeast.

A continuation of the trends described above is likely to happen even if there is no change in the locational pattern of manufacturing industry as a result of structural changes in energy prices. But it is not likely that the locational pattern of industry will be undisturbed by such price changes. Historically, there have been relatively few energy-intensive manufacturing processes. Aluminum reduction and various other electrolytic processes have located where energy prices are low. But the present locational pattern of industry was determined at a time when *all forms* of energy were available at relatively low prices. Until 1969, energy prices tended to fall in relation to other prices [21, Chart 1]. But now that the era of cheap energy is behind us, energy could become a much more important locational determinent in the future. Some manufacturers might be forced to seek sites where energy costs are relatively low in order to remain competitive.

Most forecasts of energy supply indicate rapid depletion of the nation's stock of natural gas. This stock will be supplemented by synthetic gas made from coal. High-Btu gas would be a direct substitute for today's natural gas, and would be marketed in the same way. But an important development on the technological horizon is the potential production of low-Btu gas. This gas would have to be burned at or close to the source of production because of its low heat content in relation to volume. But the development of successful low-Btu gasification processes—and these are likely within the next five to ten years—could provide an abundant supply of clean and relatively low-cost energy in or near the nation's coal fields. There probably will not be a mass relocation of industry—such as the North-South relocation of the textile industries—as a result of low-

Btu gasification, but under present circumstances the relocation of even a much smaller number of major plants could have serious consequences for the economy of the Northeast. The endurance of the Northeast's manufacturing sectors, which so impressed Perloff and his associates, may be threatened.

The Northeast already has a large service sector, but it is likely to become even more dependent on services for employment and income in the future than it has in the past. The service sectors benefit less from productivity gains than materials-processing sectors. Also, the prices of goods are likely to rise faster than the prices of services —again because of relative energy content—and if this happens it could contribute to a further worsening of the terms of trade between the Northeast and the nation's energy-producing states.

WILL THE NORTHEAST BECOME
AN ECONOMIC DISASTER AREA?

If the Northeast's terms of trade are seriously altered by rising energy prices, and by a shift of real income from the Northeast to energy-producing states, will this region become a second Appalachia? Presumably, this could happen, but in my view it is not likely to happen. Still, the region is faced with serious economic problems; it is at a serious disadvantage when compared with other regions in the nation today. This means that adjustments will have to be made.

The industrialized nations of the world—and this means largely the industrialized *regions* of these nations—have been on a growth binge for more than a quarter of a century.[13] It would not have been possible for the trends in *per capita* energy consumption of the past quarter century to continue without some massive technological breakthrough such as successful nuclear fusion or the development of solar energy storage and transmission. Even the most optimistic of scientists and engineers working on these processes do not see them as early possibilities. Some of the proposals that have been made to date are not technologically feasible, and some involve what the physicist P.W. Bridgeman and Georgescu-Roegen have called "entropy boot-legging"—they violate known scientific laws.

But this conclusion is not message of despair. As is true of most prolonged binges, the energy binge of the past quarter century is being followed by a hangover. Hangovers hurt, but people rarely die from them. The nation has learned to adjust to other structural changes in the past, and we will learn to adjust to high energy prices in the future.

One direction of change is that suggested by E.F. Schumacher in his book *Small is Beautiful, or Economics as if People Mattered*

[31] . Since the Industrial Revolution the United States, and other Western nations, have been striving for—and achieving—more and more industrial output. This has been accomplished by the substitution of physical capital and vast amounts of energy for human effort. But as resources and energy become increasingly scarce in relation to the demand for them—and this demand will be stimulated by a growing population—the earlier trend will have to be reversed at some point. We will have to learn to use energy more wisely; for example, fewer automobiles and more bicycles. If we do not conserve energy voluntarily, the market will do it for us by way of the price system. But if we rely entirely on the market to allocate energy, there are almost certain to be many inequities and perhaps more than a few incongruous results.[14]

The most difficult task faced by the nation's leaders at all levels of government will be that of helping to change attitudes. This job will not be made any easier by the difference in views within the intellectual community. The neo-Malthusians and the neo-Victorians could not be farther apart in their assessments of the future. Perhaps our Senators and Congressmen can thus be forgiven for their timorous behavior with respect to a national energy policy. What are they to believe? Is there really an "energy crisis" or are we merely the victims of the machinations of a gigantic domestic and international petroleum cartel? Could we solve the problem by legislation? This depends, of course, on what "problem" they are talking about.

As I read the signals that are coming out of Washington, our legislators imply that they are trying to find ways of return to cheap energy. The technological optimists tell them that this can be done, but that they do not tell them how soon. Perhaps, however, the members of Congress should pay more attention to the reports issued by their own Office of Technological Assessment (OTA). OTA was established precisely to help Senators and Congressmen understand highly technical issues on which they have to legislate. And the reports issued by OTA on the energy crisis which have been made public clearly do not suggest an early or simple solution to the problem of high energy prices.

A growing number of geological studies lead to the conclusion that the world is not far from the end of the petroleum era. Even with continued technological progress in the energy field—and this progress is essential to the survival of industrial society—there will be no return to cheap energy. This will not mean disaster for the Northeast, but it does mean that adjustments will have to be made. The sooner the *need* for adjustment is recognized—and the earlier legislative leaders plan for the changes ahead—the more orderly the adjustment process can be.

NOTES TO CHAPTER 3

1. For purposes of this study the Northeast is defined as the six New England states plus the three Middle Atlantic states: New York, New Jersey, and Pennsylvania. Later in the paper, however, Pennsylvania is "shifted" to the South Atlantic region when certain comparisons are made. The reasons for doing this will be obvious from the discussion in the text.

2. In nontechnical terms this means that the value of a "bundle" of goods and services produced for export to other regions was greater than the value of the bundle of goods and services imported.

3. This led to some unwarranted complacency in the region because it was widely believed that the displaced textile workers were finding jobs in the expanding electronics firms. An empirical study of six major textiles producing centers [19] showed that this was essentially not the case, and that the manufacturing base of the Northeast was weakened by the loss of its textile mills.

4. The relevant literature is cited in [19], p. 1.

5. For a graphic illustration see [6], p. 15.

6. I am indebted to my colleague John Stasney for a stimulating discussion of this issue, and for directing me to much of the literature discussed in the first part of this section.

7. Not all economists, let me hasten to add, believe in the ubiquity of price competition, and the inability of sellers to get together in order to fix prices for their mutual benefit. Indeed, many economists are fond of quoting a long passage from Adam Smith's *Wealth of Nations* to the effect that whenever they have an opportunity to do so, businessmen will conspire to fix prices.

8. For further discussion on this point see [21].

9. See, for example, [36], p. 388.

10. The classification of states by region is as follows:

New England: Maine, Massachusetts, Connecticut, Vermont, New Hampshire, and Rhode Island.

Middle Atlantic: New York, New Jersey, and Pennsylvania.

East North Central: Ohio, Indiana, Illinois, Michigan, and Wisconsin.

West North Central: Minnesota, Iowa, Missouri, North Dakota, South Dakota, Nebraska, and Kansas.

South Atlantic: Delaware, Maryland and D.C., Virginia, West Virginia, North Carolina, South Carolina, Georgia, and Florida.

East South Central: Kentucky, Tennessee, Alabama, and Mississippi.

West South Central: Arkansas, Louisiana, Oklahoma, and Texas.

Mountain: Colorado, Utah, Montana, Idaho, Wyoming, New Mexico, Arizona, and Nevada.

Pacific: Washington, Oregon, California, and Alaska (data for Hawaii not included in this study).

11. In this context the term "exports" refers to other regions as well as other nations.

12. For a good discussion of the principles involved, see [13], pp. 94–98.

13. I realize that throughout our history the American economy has experienced growth, punctuated by business cycles, but the post-World War II expansion has no historical parallel.

14. As an example, Georgescu has pointed out the incongruity of the mechanical golf cart, which not only uses energy, but also takes the exercise out of an athletic event [11].

REFERENCES

1. "Adjusting to Scarcity." *Annals of the American Academy of Political and Social Science.* Philadelphia: AAPSS, July 1975.
2. Adleman, M. A. "Population Growth and Oil Resources." *Quarterly Journal of Economics* 89 (May 1975): 271–75.
3. American Association for the Advancement of Science. *Science* 184 (April 1974). This issue is devoted to the subject of energy.
4. Bright, Arthur A., Jr., et al. *The Economic State of New England.* New Haven: Yale University Press, 1954.
5. Buckley, Jerome Hamilton. *The Triumph of Time.* Cambridge, Massachusetts: Harvard University Press, 1966.
6. Committee for Economic Development. *International Economic Consequences of High-Priced Energy.* New York: CED, 1975.
7. Dickens, Charles. *Our Mutual Friend.* London: Harper & Brothers, 1865.
8. Ford Foundation. Energy Policy Project. *Exploring Energy Choices.* Washington: Ford Foundation, 1974.
9. Friedland, Edward; Paul Seabury; and Aaron Wildavsky. *The Great Detente Disaster.* New York: Basic Books, 1975.
10. Georgescu-Roegen, Nicholas. *The Entropy Law and the Economic Process.* Cambridge, Massachusetts: Harvard University Press, 1971.
11. ____. "Energy and Economic Myths." *Southern Economic Journal* 41 (January 1975): 347–81.
12. Gordon, Richard L. *Coal and the Electric Power Industry.* Baltimore: Johns Hopkins University Press, 1975.
13. Harris, Seymour. *The Economics of New England.* Cambridge, Massachusetts: Harvard University Press, 1952.
14. Hausman, Leonard J. "Economic Development in New England: A Discussion." *Labor Law Journal* 26 (August 1975): 524–27.
15. Hudson, Edward A., and Dale W. Jorgenson. "U.S. Energy Policy and Economic Growth, 1975–2000." *Bell Journal of Economics and Management Science* (Autumn 1974): 461–514.
16. Mancke, Richard B. *The Failure of U.S. Energy Policy.* New York: Columbia University Press, 1974.
17. Meadows, Donella H., et al. *The Limits to Growth.* New York: Universe Books, 1972.
18. Meadows, Dennis L., et al. *Dynamics of Growth in a Finite World.* Cambridge, Massachusetts: Write-Allen, 1974.
19. Miernyk, William H. *Inter-Industry Labor Mobility.* Boston: Bureau of Business and Economic Research, Northeastern University, 1955.
20. ____. "Regional Employment Impacts of Rising Energy Prices." *Labor Law Journal* 26 (August 1975): 518–23.
21. ____. "The Regional Economic Consequences of High Energy Prices." *Economic Development—USA*, Vol. 1, No. 1 (forthcoming).

22. Miernyk, William H. "Some Regional Impacts of the Rising Cost of Energy." Prepared for the Annual Meeting, Regional Science Association, Cambridge, Massachusetts, November 7–9, 1975.
23. National Science Foundation, *Energy, Environment, Productivity.* Proceedings of the First Symposium on Research Applied to National Needs. Washington: GPO, 1973.
24. Office of Technology Assessment. *An Analysis Identifying Issues in the Fiscal Year 1976 ERDA Budget.* OTA Serial D. Washington: GPO, March 1975.
25. Olson, Macur, and Hans W. Landsberg. *The No-Growth Society.* New York: Norton, 1973.
26. Perloff, Harvey S., et al. *Regions, Resources, and Economic Growth.* Baltimore: Johns Hopkins University Press, 1960.
27. Polenske, Karen R., and Paul F. Levy. *Multiregional Economic Impacts of Energy and Transportation Policies.* Washington: U.S. Department of Transportation, DOT Report No. 8, March 1975.
28. Rand, Christopher T. *Making Democracy Safe for Oil.* Boston: Little, Brown, 1975.
29. Randall, John H., Jr. *The Making of the Modern Mind.* Boston: Houghton-Mifflin, 1926.
30. Roe, Frederick William. *Victorian Prose.* New York: Ronald Press, 1947.
31. Schumacher, E. F. *Small is Beautiful.* New York: Harper & Row, 1973.
32. *Scientific American* (September 1971). This issue is devoted to the subject of Energy and Power.
33. Solow, Robert M. "The Economics of Resources or the Resources of Economics." *American Economic Review* 64 (May 1974): 1–14.
34. Stahl, Sheldon W. "On Economic Growth." *Monthly Review* (Federal Reserve Bank of Kansas City) February 1975, pp. 3–9.
35. *Summary Report of the Cornell Workshop on Energy and the Environment.* U.S. Senate Committee on Interior and Insular Affairs, Serial No. 92–23. Washington: GPO, May 1972.
36. Weinstein, Milton C., and Richard J. Zeckhauser. "Optimal Consumption of Depletable Resources." *Quarterly Journal of Economics* 89 (August 1975): 371–92.

Chapter 4

Some Regional Impacts of the Rising Costs of Energy

William H. Miernyk

Few events have had as dramatic an impact on the American economy as the structural changes in world export prices triggered by the Organization of Petroleum Exporting Countries (OPEC) cartel in 1973. The sudden quadrupling of oil prices produced a spate of books and journal articles dealing with various aspects of the "energy crisis." Quite naturally, these publications reflect their authors' biases and predilections.

Some writers have concluded that there is nothing to worry about. They have argued that OPEC contains the seeds of its own destruction. Like all cartels, they insist, the OPEC organization is inherently unstable. High petroleum prices will lead to a temporary glut; this in turn will lead to price shading, then outright price cutting. Soon after this the whole elaborate structure of administered petroleum prices will come tumbling down. The result will be a return to the cheap energy that has been taken for granted throughout most of the history of the United States.[1] This train of thought has been discussed in some detail in an earlier paper [12].

There is a backup argument for those unwilling to accept the assumption that OPEC is inherently unstable. In this view technology will "solve" the energy problem as it has supposedly solved other economic problems in the past. A somewhat extreme example of the revival of an early Victorian belief in unlimited technological and

Reprinted from *Papers of the Regional Science Association*, Vol. 37 (1976). Part of this paper is based on work being done in collaboration with Frank Giarratani. The research assistance of Melissa Wolford and Charles Socher is gratefully acknowledged.

scientific progress is given in a recent article by Weinstein and Zeckhauser [21]. After making the customary assumptions about a perfectly competitive market, Weinstein and Zeckhauser go through the
traditional marginal analysis and—not surprisingly—conclude that
"underlying market forces will work to produce appropriate rates of
resource consumption" [21, p. 390]. They find this result to be a
"powerful solace." And while nothing in their analysis leads to their
ultimate conclusion, Weinstein and Zeckhauser end their article with
a 1955 quotation from John von Neumann: ". . . A few decades hence
energy may be free—just like the unmetered air—with coal and oil
used mainly as raw materials for organic chemical synthesis, for
which, as experience has shown, their properties are best suited."

The conclusions reached by Weinstein and Zeckhauser are not
exceptional in the burgeoning post-1973 literature on the energy
crisis. Indeed, the vanguard of technological optimists appears to
be made up of economists. Wilfred Malenbaum, whose work on economic development is well known, has recently argued that human
inadequacy—not resource constraints—is the root cause of today's
economic difficulties [10]. The list of others who argue in a similar
vein could easily be extended. In the same volume which includes
Malenbaum's paper, however, there is an article by Ian D. MacGregor,
a geologist, which gives a more sobering view of resource supply constraints [9]. The neo-Victorian optimism so prevalent among economists is notable by its absence in the writings of many geologists
and other earth scientists.

One of the characteristics of those studies of the energy crisis
which conclude on an optimistic note is a tendency to ignore or gloss
over the supply side of the problem. In many cases there is no explicit
mention of the supply of energy resources; in others there is an
implicit assumption that the supply is virtually inexhaustible. One
of the few economists to challenge the neo-Victorian technological
optimists has been Nicholas Georgescu-Roegen [5]. Georgescu does
not accept the extreme view, which has been advanced by Dennis
and Donella Meadows [11], that a Malthusian catastrophe is immi-
nent. But he also rejects the unwarranted optimism of economists
and others who ignore energy supply constraints, and who assume
that science and technology will find ways to provide virtually unlimited quantities of energy. Georgescu has buttressed his economic
analysis by introducing the basic principles of thermodynamics. And
he feels that many of those who engage in wishful thinking about a
future of abundant energy supplies and low energy prices are indulg-
ing in what the physicist P. W. Bridgman called "entropy bootlegging"
[5, p. 359].

While the output of books and articles dealing with the energy crisis has been increasing geometrically, little attention has been paid to the regional consequences of high and rising energy prices. This article is a progress report on the early stages of a study which is attempting to look into this matter in considerable detail. Because it is a progress report, it is longer on hypotheses than on either empirical findings or methodological innovations.

A basic assumption of the present study is that Georgescu-Roegen is correct, that there is a *worldwide* shortage of energy, and that the shortage will not be alleviated by an early and magical technological cure. Georgescu and those who share his views recognize the long-range potential of nuclear fusion and solar energy, but there is nothing on the horizon to suggest that either of these will make significant contributions to energy supplies for the remainder of this century. While there will certainly be an increase in the number of conventional nuclear power plants, most of the increased demand for energy during the next fifteen to twenty years will be satisfied by increasing the production of fossil fuels. And substantial increases in output will lead to rising marginal costs for coal, petroleum, and natural gas.

THE PRODUCTION AND CONSUMPTION OF BASIC ENERGY, BY REGION

The production and consumption of basic energy resources by census region are given in Table 4−1.[2] Similar tables have been prepared on a state basis, and the larger tables also show the changes in production-consumption patterns that have occurred since the early 1960s. These tables are far too cumbersome to be included here, however, and much of the interesting detail is regrettably lost in the process of aggregation. Nevertheless, Table 4−1 shows which of the census regions are energy consumers (ECs) and which are energy producers (EPs). Although the basic data leave something to be desired, Table 4−1 shows which regions are likely to gain and which are likely to lose as the result of rising energy prices.[3]

New England, the Middle Atlantic states, the East and West North Central states, and the Pacific Coast are net energy consumers. All consume more of the three energy resources included in Table 4−1 than they produce. New England stands out in particular because it produces none of these energy resources—it is a *pure* energy consumer. New England is also the only region which purchases no domestic crude petroleum. This region has done more than others to reduce its dependence on domestic coal since 1965. It had become entirely dependent upon foreign suppliers of crude and residual fuel

Table 4–1.   Production and Consumption of Basic Energy Resources, by Region, 1972

Energy Source and Region	Production	Consumption	Net Imports (−) or Exports (+)	Net Imports or Exports ÷ Population
Coal (net tons × 10³)				
New England . . .	0	1,522	−1,522	−0.13
Middle Atlantic . .	75,939	78,998	−3,059	−0.08
East North Central .	142,439	206,504	−64,065	−1.59
West North Central .	13,261	39,587	−26,326	−1.61
South Atlantic. . .	159,411	96,907	+62,504	+2.0
East South Central .	153,262	78,843	+74,419	+5.77
West South Central .	7,097	930	+6,167	+0.32
Mountain . . .	40,675	26,330	+14,345	+1.73
Pacific	3,302	5,352	−2,050	−0.08
U.S. Total . . .	595,386	534,973	+60,413	+0.29
Domestic Crude Petroleum (bbls × 10³)				
New England . . .	0	0	0	0
Middle Atlantic . .	4,459	4,561	−102	−0.003
East North Central .	63,352	66,971	−3,619	−0.090
West North Central .	103,352	105,816	−2,464	−0.151
South Atlantic. . .	19,574	19,135	+439	+0.014
East South Central .	80,934	81,469	−535	−0.041
West South Central .	2,419,664	2,425,370	−5,076	−0.263
Mountain . . .	344,118	348,510	−4,392	−0.530
Pacific	419,915	420,600	−685	−0.028
U.S. Total . . .	3,455,368	3,472,432	−17,064	−0.084

Natural Gas (cu. ft. × 10^6)

New England	0	260,049	−260,049	−21,959.9
Middle Atlantic	77,637	1,843,049	−1,765,412	−47,470.1
East North Central	125,765	4,132,896	−4,007,130	−99,551.1
West North Central	925,227	2,088,836	−1,163,609	−71,303.9
South Atlantic	233,503	1,533,501	−1,299,998	−42,385.2
East South Central	171,306	1,196,300	−1,024,994	−79,500.0
West South Central	18,603,927	8,024,175	+10,579,752	+547,577.9
Mountain	1,781,459	1,364,726	+416,733	+50,317.9
Pacific	612,874	2,565,914	−1,953,040	−80,757.5
U.S. Total	22,531,696	23,009,445	−447,747	−2,215.0

Source: *Minerals Industry Yearbook*, 1972.

oils, and was thus particularly vulnerable to the 1973 increases in world petroleum prices (cf. [17, p. 18]; [6]).

None of the census regions is a net exporter of all of the domestic energy resources listed in Table 4−1. The South Atlantic, the two South Central, and the Mountain regions are net exporters of coal. The West North Central region is the only net exporter of domestically produced petroleum, and this by a slim margin. The West South Central region—which includes Texas and Louisiana—and the Mountain regions are net exporters of natural gas. On the basis of this preliminary view of energy production-consumption relationships, one would conclude that the South Atlantic, the South Central, and the Mountain regions would be the most likely regional beneficiaries of windfall gains due to energy price increases.

Table 4−1 deals with the quantities of energy produced and consumed by regions, whereas the dramatic changes that have occurred in recent years have been in prices. Thus, this table does not show which regions might have benefited and which might have been penalized by their relative positions as energy producers or energy consumers.

THE REGIONAL IMPACT OF RISING ENERGY PRICES

The U.S. input-output model has been disaggregated on a regional basis by Leontief and Bezdek, and the modified models have been used to analyze the regional impacts on production and employment of assumed cuts in defense spending [7, 1]. Polenske also has constructed a disaggregated national model—the Multiregional Input-Output Model (MRIO)—and has recently used it to study the regional impacts of possible changes in energy and transportation policies [16]. Earlier, Leontief showed how a national input-output model could be used to trace the direct and indirect effects of price changes on the industrial sectors of an economy [8]. In principle, although methodologically much simpler, the present analysis is related to these earlier studies. It is worth repeating that the results given here are preliminary; they should be considered as only rough approximations.

Polenske's report contains much useful information, and provides an excellent illustration of how the MRIO model can be used for the analysis of policy alternatives. It is primarily concerned with the regional impacts of alternative electricity-generating technologies on the use of coal, and thus on the transportation network. Her study does not deal with price changes, however, and shows only some of the effects of changes in interregional coal shipments.[4]

The first step in the present study was to recalculate the 1967 national input-output table in terms of the latest available prices (1974) [18]. This was done by multiplying each row of the transactions table T_i by an appropriate price index P_i. For this purpose the detailed price indexes published by the Bureau of Labor Statistics [19, 20] were used. All row entries in the 1967 transactions table were adjusted for price changes, including the final demand sectors and total gross output.

Following the price adjustments, a new input coefficient matrix (A^*) was calculated. The coefficients in A^* reflect changes in *relative* prices. Finally, a new Leontief inverse was calculated from the price-adjusted data, that is, $(I - A^*)^{-1}$. This shows the direct and indirect effects on each sector of price changes between 1967 and 1974.

Since the entire transactions table was adjusted for price changes it was possible to identify the impacts on intermediate inputs. These relative price changes are given in Figure 4–1. There is relatively little variation in price changes of intermediate inputs, except for sectors 7 (coal mining), and 31 (petroleum refining and related industries). The relatively small variation in price changes for intermediate goods suggests that the regional effects of such price changes were relatively small. Of greater interest, however, are the changes in *value added* engendered by price changes during the period covered by this study. Changes in value added, while not limited to windfall profits, would include such windfalls when they occurred.

Value added was calculated as a residual from the following relationship:

$$X_j^* - \sum_{j=1}^{n} x_j^* = VA_j^*, \ (n = 78)$$

in which the asterisks indicate that all elements of the input-output table have been adjusted for price changes. X_j^* = total gross outlays by sector; Σx_j^* is the column sum of intermediate transactions, and the difference between these two is, of course, value added by sector.

The results of these calculations are summarized in Figure 4–2, which shows wide variation in value added by sector as a result of price changes between 1967 and 1974.[5] The increase in value added in sector 7 (coal mining) is about 454 percent; this is followed by sector 31 (petroleum refining and related industries), which registered an increase of almost 220 percent. Sector 8 (crude petroleum and

Figure 4–1. Relative Impacts of Price Changes on Intermediate Inputs *(percent change, 1967–1974)*

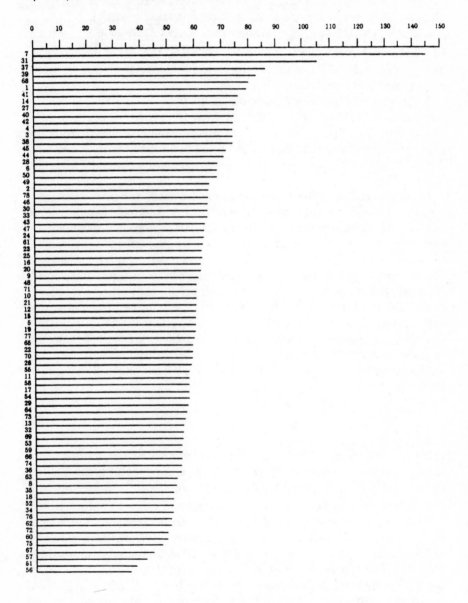

Figure 4–2. Relative Impacts of Price Changes on Value Added
(percent change, 1967–1974)

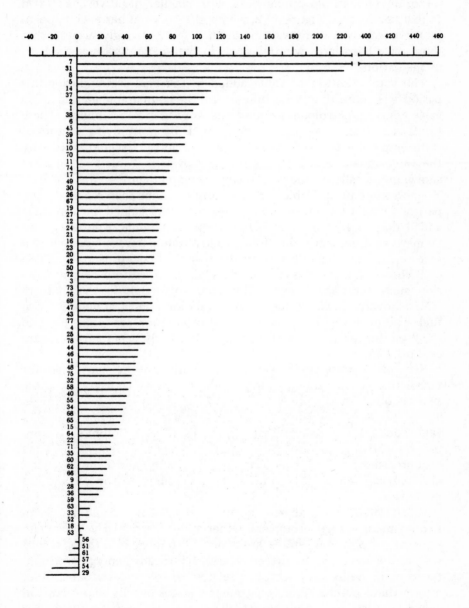

natural gas) ranked third with an increase in value added of almost 160 percent.

At the other end of the scale there were sizable negative changes in value added in sector 29 (drugs, cleaning and toilet preparations), and sector 54 (household appliances), with smaller negative changes in four other sectors. Figure 4−2, which reflects what has been happening to prices on a worldwide basis since 1967, strongly suggests that there have been large windfall gains in the energy-producing sectors of the nation.[6]

The total addition to value added in the coal sector during this period was more than $8.5 billion, and the comparable increment to value added in petroleum refining was more than $15.7 billion. These totals were distributed on a regional basis by multiplying vectors of state employment coefficients in the coal and petroleum sectors by the value-added totals. The results are given in Table 4−2, and are shown graphically on the accompanying map (Figure 4−3).[7]

The state data in Table 4−2 are more revealing than regional comparisons would have been. The states that have gained most in value added due to coal price increases are West Virginia, Kentucky, and Pennsylvania, in that order. Illinois and Virginia have substantial coal sectors and have benefited from the revival of coal prosperity. Table 4−2 shows that despite the publicity given to western coal production, such states as Colorado, Montana, North Dakota, and Utah are still relatively small producers of the nation's most abundant fossil fuel. The most rapid *increases* in coal production are taking place in the West, however, and western states will clearly benefit from future coal price increases.

Separate employment coefficients could not be calculated for petroleum and natural gas, and it was necessary to aggregate these two "goods" for the present analysis. The results show few surprises. Texas, Louisiana, Oklahoma, and California are the leading gainers in value added due to rising oil and gas prices. Kansas, Colorado, Wyoming, and Mississippi also show significant gains. Three of the large coal-producing states—West Virginia, Kentucky, and Pennsylvania— also participated in the windfall gains accruing to the oil and gas industry.

As of 1973, fewer than one-fourth of all states could be classified as major energy producers. Other states, particularly Alaska and some of the Mountain states, are expected to move into this category within the next ten to fifteen years. Historically, energy prices in the United States have been low relative to the prices of most other intermediate goods. As a consequence, states heavily dependent on energy sectors for employment—West Virginia and Kentucky, for

**Table 4-2. Employment Coefficients and Value Added, by State,
1967-1974**

	Coal		Petroleum & Natural Gas	
State	*Employment Coefficients (1973)[a]*	*Value Added (000)*	*Employment Coefficients (1973)[a]*	*Value Added (000)*
Wyoming	0.28	$ 23,839	2.95	$ 464,418
West Virginia	31.44	2,676,802	1.33	209,382
Louisiana	0	—	19.44	3,060,439
New Mexico	.59	50,233	3.25	511,647
Alaska	.07	5,960	.56	88,161
North Dakota	.21	17,879	.64	100,755
Oklahoma	.35	29,799	8.79	1,383,810
Texas	.59	50,233	36.51	5,747,769
Kentucky	15.63	1,330,738	1.09	171,599
Utah	1.09	92,803	.99	155,856
Montana	.08	6,811	.60	94,458
Kansas	.21	17,879	3.18	500,627
Colorado	.67	57,044	2.39	376,258
Virginia	8.83	751,786	.05	7,871
Pennsylvania	20.20	1,719,828	1.32	207,808
Mississippi	0	—	2.05	322,732
Alabama	3.39	288,625	.21	33,060
Nevada	0	—	.09	14,169
California	0	—	7.89	1,242,123
Illinois	7.23	615,562	1.72	270,780
All others	9.14	778,179	4.95	779,278
Total	—	$8,514,000	—	$15,743,000

[a] The most recent employment data available at the time this was written were for the year 1973. While employment coefficients change over time, the changes occur slowly. It is not likely that the year-to-year change between 1973 and 1974 would have been large enough to introduce any appreciable error in the distribution of value added.

example—have been at the low end of the regional distribution of per capita income (cf. [13]). Even those states noted for their oil and gas fortunes—Texas, Oklahoma, and Louisiana—have been below the national average. If the terms of trade continue to turn in favor of energy-producing states, however, it is likely that there will be some convergence between the nation's energy-producing and energy-consuming states in per capita incomes.

There are limitations to the procedure that has been followed thus far. First a single price is assumed for each of the two energy "goods" included in this study. In fact, there are wide variations in mine-mouth coal prices, as well as in oil and gas prices at the wellhead. In the case of coal, for example, the assumption of a single price understates the proportions of total value added that have gone to states such as Alabama, Pennsylvania, Virginia, and West Virginia, where

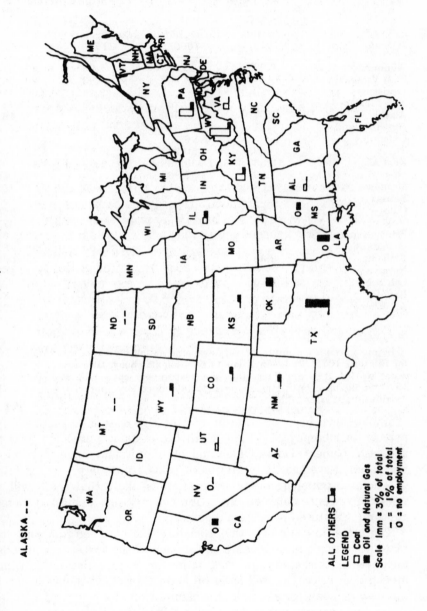

Figure 4–3.   Distribution of Incremental Value Added, by State, 1967–1974

the price of coal in 1973 was well above national and regional averages. And it has overstated the amounts flowing to Kentucky, Maryland, Ohio, and Tennessee, where the average price of coal was below national and regional averages. The same is no doubt true in the case of petroleum and natural gas. Because of the aggregation of these two "goods," no effort was made to calculate state prices. In subsequent work, however, it should be possible to take state price differentials into account when allocating the effects of price changes to places of production.

Up to this point the study shows only the states that have benefited from increases in value added attributable to rising energy prices. From a *regional* point of view, however, this is a zero-sum game—what some states gain others must lose. Thus Table 4–2 and the map in Figure 4–3 show only the positive regional impact of changing energy prices between 1967 and 1974.[8] To estimate the negative impacts—that is, the regional incidence of higher energy costs—it will be necessary to identify the origins and destinations of energy shipments, to estimate the transfer costs involved, and to adjust for regional price differentials.

Some tentative conclusions can be drawn from this preliminary analysis despite the limitations mentioned above. First, consumers, no matter where located, are feeling the effects of higher energy prices. In addition, energy-consuming regions—such as New England —are experiencing a shift in their interregional terms of trade; they have to pay more for the energy they import in terms of the goods and services they export to other regions. One might argue that after an appropriate time lag other prices will catch up with energy prices so that the pre–1973 relationship will be reestablished. But this argument is nothing more than a variation of the views expressed by technological optimists who believe that in the long run energy prices will *fall* to reestablish earlier relationships.

Energy specialists tend to believe that this is not likely to happen. Those in charge of the nation's energy research program appear to believe that for the immediate future—the next fifteen to twenty years—most of the growing demand for energy will have to be satisfied by increasing the production of fossil fuels. And even if one makes fairly optimistic assumptions about increasing productivity, the cost of fossil fuels is likely to rise more rapidly than the general price level. Thus the changes in regional terms of trade are likely to result in shifts in *real* per capita income among regions. Energy-producing regions are likely to gain at the expense of energy-consuming regions. While there are still some dissenters, such as Gordon [4], their arguments about the deficiencies of coal as a fuel are far from convincing. Consensus is growing among energy forecasters that coal will account for an increasing, rather than a declining, proportion of future energy supply.[9]

Other environmental concerns, which appear to have taken a back seat to the more pressing concerns of energy supply and energy prices, have not been forgotten. The Energy Research and Development Administration, for example, is trying to achieve the best of both worlds. It has shifted research funds from long-range nuclear projects to coal liquefaction and gasification projects which could have an earlier payoff in terms of providing clean fuel from the nation's most abundant energy resource.

The direction and success of various facets of the coal liquefaction and gasification programs could have further regional impacts. Liquefied coal would supplement domestically produced and imported crude and residual fuel oils. High-Btu gas would supplement the stock of natural gas produced domestically and imported in liquid form from abroad. Neither of these should have significant locational impacts, although they could contribute to further shifts in the terms of interregional trade. The successful development of low-Btu gas at competitive prices could, however, have an impact on the location of industry. Low-Btu gas is suitable primarily for use as a boiler fuel. It cannot be shipped long distances [15]. But the availability of substantial quantities of low-Btu gas could attract various energy-intensive production activities to the coal fields where this gas would be produced and could result in a significant amount of industrial relocation similar to the North-South shift of the textile industries.

In its study of the international economic consequences of high-energy prices, the Committee for Economic Development has stated that: "The shocks and repercussions produced by sharply rising international energy costs have placed great strains on the world's market economies" [2, p. 7]. While the world appears to be adjusting to the new structure of energy prices, it is clear that the terms of trade between the OPEC nations and the rest of the world have been altered. There is still a tendency, however, to overlook the regional impacts of such drastic structural changes. But there is no reason to believe that the differential impacts on energy-producing and energy-consuming regions are less significant—or less permanent—than those that have recently altered the relationship among energy-producing and energy-consuming nations.

NOTES TO CHAPTER 4

1. To avoid ambiguity, let me stress that the term "cheap energy" in the text means that energy prices were low *relative to* other prices. Until the early 1970s, the ratios of wholesale fuel prices to all commodity prices in the United States either were stable or declining. Since 1973 these ratios have skyrocketed. On this see [12], Chart 1.

2. *New England*: Maine, Massachusetts, Connecticut, Vermont, New Hampshire, and Rhode Island. *Middle Atlantic*: New York, New Jersey, and Pennsylvania. *East North Central*: Ohio, Indiana, Illinois, Michigan, and Wisconsin. *West North Central*: Minnesota, Iowa, Missouri, North Dakota, South Dakota, Nebraska, and Kansas. *South Atlantic*: Delaware, Maryland and D.C., Virginia, West Virginia, North Carolina, South Carolina, Georgia, and Florida. *East South Central*: Kentucky, Tennessee, Alabama, and Mississippi. *West South Central*: Arkansas, Louisiana, Oklahoma, and Texas. *Mountain*: Colorado, Utah, Montana, Idaho, Wyoming, New Mexico, Arizona, and Nevada. *Pacific*: Washington, Oregon, California, and Alaska (data for Hawaii not included in this study).

3. Production and consumption figures do not balance in the case of coal because there are losses in transit, due primarily to the evaporation of water. The relatively small difference between total production and consumption in the case of petroleum is explained largely by inventory changes and reporting error. The same explanation also, no doubt, applies to natural gas.

4. For this prototype study Polenske restricted data adjustments to sector 68 (electric, gas, water, and sanitary services) and to the trade flows of coal. Her model could easily be broadened, however, to reflect changes in input structures in all sectors based on whatever assumptions one wished to make about the supplies and prices of alternative fuels.

5. The names of the sectors are given in the appendix.

6. For an excellent summary of the relative constancy of energy prices and other world export prices between 1955 and 1971, and the explosive divergence between these sets of prices following 1971, see [2], p. 15.

7. The rank of sectors by value added change is given in the appendix. There is little correlation between the rankings of sectors by intermediate price changes and by changes in value added (the Spearman coefficient is 0.458).

8. Compare, for example, Leontief's map showing the positive and negative regional impacts of *compensated* cuts in defense spending in [7], p. 225.

9. See, for example, [3].

APPENDIX

Sector	Rank of Price Effects	
	On Intermediate Inputs	On Value Added
7 Coal mining	1	1
31 Petroleum refining and related industries .	2	2
37 Primary iron and steel manufacturing .	3	6
39 Metal containers	4	12
68 Electric, gas, water, and sanitary services .	5	54
1 Livestock and livestock products . .	6	8
41 Stampings, screw machine products and bolts .	7	46
14 Food and kindred products	8	5
27 Chemicals and selected chemical products .	9	24
40 Heating, plumbing, and structural metal products .	10	51
42 Other fabricated metal products . .	11	31
4 Agricultural, forestry, and fishery services .	12	41
3 Forestry and fishery products . . .	13	34
38 Primary nonferrous metal manufacturing .	14	9
45 Construction, mining and oil field machinery .	15	11
44 Farm machinery and equipment . . .	16	44
28 Plastics and synthetic materials . .	17	65
6 Nonferrous metal-ores mining . . .	18	10
50 Machine shop products	19	32
49 General industrial machinery and equipment .	20	19
2 Other agricultural products . . .	21	7
78 State and local government enterprises .	22	43
46 Materials handling machinery and equipment .	23	45
30 Paints and allied products	24	20
33 Leather tanning and industrial leather products .	25	69
43 Engines and turbines	26	39
47 Metalworking machinery and equipment .	27	38
24 Paper and allied products; except containers .	28	26

29	61	Other transportation equipment	75
30	23	Other furniture and fixtures	29
31	25	Paperboard containers and boxes	42
32	16	Broad and narrow fabrics, yarn and thread mills	28
33	20	Lumber and wood products; except containers	30
34	9	Stone and clay mining and quarrying	64
35	48	Special industry machinery and equipment	47
36	71	Real estate and rental	59
37	10	Chemical and fertilizer mineral mining	14
38	21	Wooden containers	27
39	12	Maintenance and repair construction	25
40	15	Tobacco manufacturers	56
41	5	Iron and ferroalloy ores mining	4
42	19	Misc. fabricated textile products	23
43	77	Federal government enterprises	40
44	65	Transportation and warehousing	55
45	22	Household furniture	58
46	70	Finance and insurance	15
47	26	Printing and publishing	21
48	55	Electric lighting and wiring equipment	52
49	11	New construction	16
50	58	Misc. electrical machinery, equipment and supplies	50
51	17	Misc. textile goods and floor coverings	18
52	54	Household appliances	77
53	29	Drugs, cleaning and toilet preparations	78
54	64	Misc. manufacturing	57
55	73	Business services	35
56	13	Ordinance and accessories	13
57	32	Rubber and misc. plastics products	49
58	69	Wholesale and retail trade	37
59	53	Electric industrial equipment and apparatus	72
60	59	Motor vehicles and equipment	67
61	66	Communications, except radio and TV broadcasting	63
62	74	Automobile repair and services	17

(Appendix continued overleaf)

Appendix *(continued)*

Sector	Rank of Price Effects	
	On Intermediate Inputs	On Value Added
36 Stone and clay products	63	66
63 Optical, ophthalmic, and photographic equipment .	64	68
8 Crude petroleum and natural gas	65	3
35 Glass and glass products	66	60
18 Apparel	67	71
52 Service industry machinery	68	70
34 Footwear and other leather products . . .	69	53
76 Medical, educational services, and nonprofit organizations .	70	36
62 Scientific and controlling instruments . . .	71	62
72 Hotels; personal and repair service; except automobile .	72	33
60 Aircraft and parts	73	61
75 Amusements	74	48
67 Radio and TV broadcasting	75	22
57 Electronic components and accessories . . .	76	76
51 Office, computing and accounting machinery . .	77	74
56 Radio, television and communication equipment . .	78	73

REFERENCES

1. Bezdek, Roger H. "The 1980 Economic Impact—Regional and Occupational—of Compensated Shifts in Defense Spending." *Journal of Regional Science* 15 (1975): 183—97.
2. Committee for Economic Development. *International Economic Consequences of High-Priced Energy.* New York: CED, 1975.
3. Federal Energy Administration. *Project Independence Report.* Washington, D.C.: Government Printing Office, 1974.
4. Gordon, Richard L. *U.S. Coal and the Electric Power Industry.* Baltimore: Johns Hopkins Press, 1975.
5. Georgescu-Roegen, Nicholas. "Energy and Economic Myths." *Southern Economic Journal* 41 (January 1975): 347—81.
6. Hausman, Leonard J. "Economic Development in New England: A Discussion." *Labor Law Journal* 26 (August 1975): 524—27.
7. Leontief, Wassily. "The Economic Impact—Industrial and Regional—of an Arms Cut." *Review of Economics and Statistics* 47 (August 1965): 217—41. Reprinted in Leontief, *Input-Output Economics.* New York: Oxford University Press, 1966, pp. 184—222.
8. _____ . "Wages, Profits, Prices and Taxes." *Dun's Review*, June 1947. Reprinted in Leontief, *Input-Output Economics.* New York: Oxford University Press, 1966, pp. 30—40.
9. MacGregor, Ian D. "Natural Distribution of Metals and Some Economic Effects." *Annals*, American Academy of Political and Social Science, Vol. 420 (July 1975): 31—45.
10. Malenbaum, Wilfred. "Scarcity: Prerequisite to Abundance." *Annals*, AAPSS, Vol. 420 (July 1975): 72—85.
11. Meadows, Donella H., et al. *The Limits to Growth.* New York: Universe Books, 1972.
12. Miernyk, William H. "The Regional Economic Consequences of High Energy Prices." Hudson Institute—EDA, August 1975.
13. _____ . "Regional Employment Impacts of Rising Energy Prices." *Labor Law Journal* 26 (August 1975): 518—23.
14. _____ ., and John T. Sears. *Air Pollution Abatement and Regional Economic Development.* Lexington, Mass.: D.C. Heath, Lexington Books, 1974.
15. Office of Technology Assessment. *An Analysis Identifying Issues in the Fiscal Year 1976 ERDA Budget.* OTA Serial D. Washington, D.C.: Government Printing Office, 1975.
16. Polenske, Karen, and Paul F. Levy. *Multiregional Economic Impacts of Energy and Transportation Policies.* DOT Report No. 8, March 1975.
17. U.S. Senate. Committee on Interior and Insular Affairs. *Summary Report of the Cornell Workshop on Energy and the Environment.* Serial No. 92—93. Washington, D.C.: Government Printing Office, 1972.
18. U.S. Department of Commerce. Interindustry Economics Division. "The Input-Output Structure of the U.S. Economy: 1967." *Survey of Current Business* 54 (February 1974): 24—56.

19. U.S. Department of Labor. Bureau of Labor Statistics. *Wholesale Prices and Price Indexes: Data for January 1975.* Washington, D.C.: Government Printing Office, 1975. Table 6, pp. 20–54; Table 11, pp. 58–60; Table 13, pp. 64–71.

20. ____. *CPI Detailed Report for January 1975.* Washington, D.C.: Government Printing Office, 1975. Table 2, p. 8; Table 8, pp. 16–19.

21. Weinstein, Milton C., and Richard J. Zeckhauser. "The Optimal Consumption of Depletable Natural Resources." *Quarterly Journal of Economics* 89 (August 1975): 371–92.

Chapter 5

Rising Energy Prices and Regional Economic Development

William H. Miernyk

Traditionally, development planners and policymakers have based their activities and programs on the export base theory embedded in a Clark-Fisher development model. The export base theory—which treats export activities as the prime movers of economic development and all other activities as supportive —is sufficiently well-known to require no elaboration. But the development concepts of Clark and Fisher belong to an earlier era, and a brief review might be helpful.[1]

An abbreviated version of the Clark-Fisher hypothesis requires oversimplification, but this need not do violence to the basic ideas. Essentially, Clark and Fisher, working independently, argued that economic development is a function of structural change. The propulsive force of economic development in the Clark-Fisher model is technological progress and the gains in productivity which new technology engenders. Technological progress in the primary sectors of agriculture, forestry, and fisheries leads to increases in output while reducing labor requirements. More labor is thus made available to the secondary sectors of manufacturing, construction, mining, and electric power production. Later, as an economy (national or regional) becomes industrialized, further technological progress in the secondary sector releases labor for employment in the trades and services which are called *tertiary activities*. Lewis Bean[2] added another dimension to the Clark-Fisher argument—with an explicit reference to regions—when he elaborated on Clark's earlier work which showed

Reprinted by permission from *Growth and Change: A Journal of Regional Development*, Vol. 8, No. 3 (July 1977): 2–7.

a close, positive correlation between the above process of industrialization and rising per capita income.[3]

Not all economists have accepted the Clark-Fisher thesis (cum Bean) uncritically. The late Seymour Harris, for example, argued that structural changes which produced gains in tertiary employment do not necessarily produce rising per capita incomes.[4] They may, in fact, reflect deterioration in a region's comparative advantages vis-à-vis the regions with which it trades. Harris's position was supported by Fourastié, who argued that a nation is not wealthy because it has a large tertiary sector; rather, cause and effect run in the other direction—only a wealthy nation can afford a large tertiary sector.[5] The same, of course, can be said of regions.

What role have energy prices played in the development process described by Clark and Fisher? Historically, energy prices in the United States have been low relative to those of other inputs. Competition among producers of oil and coal—as well as competition between these sources of basic energy—kept energy prices from rising. Until a few years ago, natural gas was so abundant relative to demand that it was often more economical to flare off the gas at the wellhead than to transport it any appreciable distance.

Cheap energy played a central part in the industrialization and urbanization of America. Since energy was abundant and cheap, the cost of transporting or transmitting energy was also low. Thus, major industrial complexes could develop in the Northeast and the Great Lakes, far removed from basic energy sources. Aluminum reduction plants, chemical and glass factories, and a few other processes were attracted to sources of cheap energy. But the locational determinants of these processes differed markedly from those of manufacturing in general. In most cases, industrial establishments were located to minimize what Hoover called *transfer costs*, that is, the cost of assembling the necessary raw materials and of shipping finished products to markets. Other plants were attracted to low-cost labor. But only in a handful of industries was energy a significant locational determinant.

The Clark-Fisher development model stands up well when used to explain the pattern of regional development in the United States between, say, the Civil War and the beginning of World War II. And when the first major effort was made to stimulate economic development in a lagging region in the United States—the Tennessee Valley Authority (TVA)—one of its objectives was that of increasing employment in the industrial sector.[6] The development policies that emerged from more recent legislation also had among their objectives the creation of industrial jobs.[7] Regional development programs in the United States have attempted to change the industrial structure of lagging regions either through investment in social overhead capital, directly productive capital, or some combination of these two.

REGIONAL IMPACT OF RISING ENERGY PRICES

It is not a simple matter to assess the regional impact of rising energy prices in the United States. This is because the quantum jump in prices came at a time when the United States was in a severe recession complicated by extreme inflation. The most immediate impact— and perhaps the only one that is entirely clear-cut up to now—has been the effect which the energy crisis has had on the fiscal positions of various states.

Unlike the federal government, states are not able to engage in deficit financing. To avoid fiscal problems during periods of recession, most states have established contingency funds. A typical state which ends a fiscal year with revenues in excess of expenditures deposits the surplus in its contingency fund. These funds are built up during good years and are drawn down during years in which expenditures exceed revenues. Thus, in the words of the Joint Economic Committee (JEC): "Shifts in the size of the unencumbered surplus are a good indication of the relative fiscal position of the states from one year to the next."[8] A survey conducted by the JEC in 1975 revealed that of the 48 states responding, 34 were drawing down their unencumbered surpluses. But the 13 states with a per capita energy input above the national average "experienced a 14-percent decline in the size of their unencumbered surplus, compared to a 52-percent decline for all other states."[9] In fact, 7 of the 13 energy states registered an increase in their unencumbered surpluses, and 1 showed no change.[10]

A similar pattern was found in 8 states that derive a significant percentage of their income from agriculture.[11] Meanwhile, the unencumbered surpluses of 18 states which are not major energy producers and which had above-average unemployment rates "experienced a combined reduction in their unencumbered surplus . . . from $2.3 to $0.4 billion, or 83 percent. This surplus is less than one percent of the combined budgets of all high unemployment states."[12]

The differential regional impact of the recession on the fiscal position of states was the result of a combination of cyclical and secular forces. All states were adversely affected by the recession. The problem was exacerbated in some states (the energy-consuming states) by rising energy prices, while the recession was cushioned in energy-producing states. In a few states—such as West Virginia and Wyoming —rising energy prices produced boom or near-boom conditions.[13]

Only 20 states can be classified as net energy producers. The remainder are, of course, net energy consumers. There is a strong inverse correlation between per capita income and energy employment in the producing areas, a correlation which is improved if four relatively small producers of energy are removed.[14] Most of the remaining net producers have per capita incomes well below the national average.[15]

Elsewhere, I have suggested that the relatively low per capita income of the nation's major net energy producers was partly a result of aggressive price competition among the producers of fossil fuels. The low energy prices engendered by price competition helped "subsidize" the rapid expansion of urban and suburban America, particularly that which took place between World War II and the early 1970s.[16] Purchasers of energy in the form of electricity, gas, and gasoline enjoyed direct consumers' surpluses; they also enjoyed indirect consumers' surpluses since the prices of manufactured products were lower than they would have been in the absence of aggressively competitive energy markets. One need not dwell on the fact that the price of imported oil is no longer determined by the forces of competition; these prices are now determined by a cartel which controls most of the oil entering world markets. And the prices of other fossil fuels tend to follow those of petroleum.[17]

The initial reaction of many economists to the energy crisis precipitated by the Organization of Petroleum Exporting Countries (OPEC) in 1973 was that it couldn't last. Their basic argument was that cartels are self-destructive, and OPEC is just another cartel.[18] If this appraisal should turn out to be correct and present high energy prices should prove to be transitory, there would be no long-run regional consequences. There is no evidence to support such cheerful optimism, however. A more reasonable assumption is that the era of cheap energy is over. It is likely that the long-run demand for energy will increase more rapidly than supply for the remainder of this century; if so, energy prices will rise more rapidly than the general price level. One consequence of this trend would be a shift of real income from energy-consuming to energy-producing states.

If the price movements hypothesized above should occur, all consumers—wherever located—would have to spend an increasing share of their budgets on energy. This would mean a decline in expenditures on other things or a decline in aggregate savings. Thus, it is not as consumers that residents of the energy-producing states would gain, but only as producers. Since the owners of coal mines and oil and gas wells do not necessarily live in the states in which basic energy is produced, there will continue to be substantial geographical dispersion of profits. But much of the windfall income generated by rising energy prices will remain in the net energy-producing areas in the form of higher wages, higher rents, and particularly in higher state and local taxes. Also, as energy prices continue to rise, there will be an incentive to invest in new basic energy facilities in the energy-producing states.

When fossil fuels were selling in price-competitive markets, state legislators were unwilling to risk placing their producers at a competitive disadvantage. But as the price of coal has increased, some legislators have recognized a potential source of new state revenue. West Virginia and Kentucky are the only states which levy significant coal taxes at present. But legislators in other states are bound to observe the favorable impacts these taxes have had on the fiscal condition of these states and follow their lead.[19]

As yet, there has been no appreciable geographical shift in income as a result of rising energy prices. But geographical per capita income differentials change at a glacial pace even when there are fairly rapid underlying changes in regional economic conditions.[20] There are also statistical difficulties involved in efforts to measure geographical shifts in real income. Although consumer price indexes exist for a number of metropolitan areas, they do not have a common base; they show only comparative rates of change over time. There are no adequate regional income deflators and thus no way to make comparisons of real income, or changes in real income, on a regional basis. Recent money income changes suggest that regional income shifts are under way. Some of the low per capita income energy producers, such as Alabama and West Virginia, have been high on the list of quarterly changes in personal income payments for the past year or two. Meanwhile, some high per capita income, industrialized energy consumers, such as Connecticut and New York, have dropped to the bottom of the list. Clearly, if these relative positions are maintained for some time, there will be shifts in income from consumers of energy to producers.

IMPACT ON THE LOCATION OF INDUSTRY

Historically, there have been few energy-oriented industries. But this situation could change if the price of energy continues to rise faster than the price of other inputs. Even within the framework of Weberian location theory, for example, as energy prices rise, energy could become a more important locational determinant than it has been. As yet, there has been no great rush to substitute coal for petroleum either in manufacturing or in electric utilities. But this has been partly because of the restraints imposed by environmental protection legislation, and partly because of inertia. If the price of imported oil rises further, however, coal liquefaction and coal gasification could become economically feasible in a relatively short time.

Low-Btu coal gasification could have a significant impact on the location of industry. Unlike high-Btu gas produced from coal, low-Btu gas must be burned at or close to the source of production. It is a clean and efficient fuel and could be made available in abundance near the nation's coalfields. While one cannot be highly specific at present about the types of manufacturing plants that could be attracted to low-Btu gas sites, any plant which would not be at a comparative transportation cost disadvantage, but which would save significantly on fuel costs, could be attracted to areas producing low-Btu gas.

Table 5–1 lists the 55 most energy-intensive input-output sectors in the American economy. Energy inputs are defined as total purchases from the following sectors: coal mining (7.00); crude petroleum and natural gas (8.00); petroleum refining and related products (31.01); electric utilities (68.01); and gas utilities (68.02).[21]

In 1967, energy inputs ranged from a high of almost 53 cents per dollar of output in gas utilities to slightly more than 2 cents per dollar of output in logging camps and logging contractors. The extremely

Table 5–1. Energy Inputs per Dollar of Output, Selected Sectors, 1967

Sector Number	Name and Location Orientation	Total Energy Input Coefficient
68.02	Gas utilities*	0.52788
31.01	Petroleum refining and related products—RT	0.52610
31.02	Paving mixture and blocks—T	0.22758
68.01	Electric utilities—U	0.15507
31.03	Asphalt felts and coatings*	0.15078
36.13	Lime—RT	0.14106
27.01	Industrial inorganic and organic chemicals*	0.14048
36.01	Cement hydraulic—RT	0.13585
36.02	Brick and structural clay tile—RT	0.12102
14.30	Manufactured ice—U	0.10353
65.06	Pipeline transportation	0.10331
36.05	Structural clay products—RT	0.09684
53.07	Carbon and graphite products*	0.08743
38.04	Primary aluminum*	0.07784
65.05	Air transportation	0.07532
36.04	Clay refactories—RT	0.06698
28.02	Synthetic rubber*	0.06244
37.01	Blast furnaces and basic steel products—T	0.05461
14.17	Wet corn milling—RT	0.05401

Table 5-1. continued

Sector Number	Name and Location Orientation	Total Energy Input Coefficient
24.06	Wallpaper and building paper and board mills*	0.05362
36.14	Gypsum products—R	0.05116
35.02	Glass containers*	0.04998
24.03	Paperboard mills*	0.04550
36.20	Mineral wool*	0.04534
65.03	Motor freight transportation and warehousing	0.04384
38.03	Primary zinc*	0.04215
37.04	Primary metal products*	0.04200
36.21	Nonclay refractories	0.03491
65.01	Railroad and related services	0.03466
36.06	Vitreous plumbing fixtures*	0.03447
24.02	Paper mills except building paper*	0.03372
36.19	Minerals, ground or treated—R	0.03220
42.04	Coating, engraving, and allied services*	0.03124
27.04	Miscellaneous chemical products*	0.03052
35.01	Glass and glass products, except containers*	0.03031
36.03	Ceramic wall and floor tile*	0.03015
36.07	Food utensils, pottery*	0.02994
28.01	Plastic materials and resins*	0.02936
36.22	Nonmetallic mineral products—R	0.02709
54.06	Sewing machines—TM	0.02671
36.09	Pottery products—RT	0.02601
65.02	Local suburban and interurban highway passenger transport	0.02553
68.03	Water and sanitary service	0.02518
37.02	Iron and steel foundries—T	0.02467
17.05	Processed textile waste*	0.02437
65.04	Water Transportation	0.02407
36.15	Cut stone and stone products—RT	0.02405
24.01	Pulp mills—RT	0.02271
36.10	Concrete block and bricks—RT	0.02234
20.02	Sawmills and planning mills, general—R	0.02185
28.03	Cellulosic manmade fibers*	0.02154
27.03	Agricultural chemicals*	0.02142
50.00	Machine shop products—U	0.02101
36.17	Asbestos products*	0.02074
20.01	Logging camps and logging contractors—R	0.02017

Source: U.S. Department of Commerce, Bureau of Economic Analysis, *Direct Requirements for Detailed Industries*, vol. 2: *Input-Output Structure of the U.S. Economy: 1967* (Washington, D.C.: Government Printing Office, 1974).

*Implies sector is not tied to either materials or markets, and not likely to be transport-oriented.

Note: Symbols following sector names have the following meanings:

    RT — resource and transport-oriented
    T — transport-oriented
    U — ubiquitous
    R — resource-oriented
    TM — export market-oriented

high energy input coefficients in gas utilities and in petroleum refining and related products reflect substantial intraindustry transactions. And this in turn is the result of aggregation, in spite of the relatively high degree of disaggregation that was achieved in the 370-order 1967 matrix.

The sectors in Table 5—1 have been classified on the basis of establishment location orientation—a classification which reflects entirely my own judgment. Establishments in sectors marked with an asterisk are not tied to either materials or markets, nor are they likely to be transport-oriented. Several of these sectors—such as chemicals, primary aluminum, and glass products—are already represented in states with relatively abundant energy supplies. New plants in other sectors could be attracted to energy sources if the cost of energy increases faster than the cost of other inputs.[22]

POLICY RESPONSES

Regional development policies have not been outstandingly successful in reducing interregional disparities in per capita income and unemployment rates, either in the United States or abroad. Two major regional development programs were started in the United States during the 1960s. Appalachian Regional Commission (ARC) programs have concentrated on investment in social overhead capital (SOC), primarily the Appalachian Development Highway System, vocational and technical educational, and health delivery facilities. It is much too early to make a definitive judgment about the success of this program, although some positive results are clearly discernible. But the revival of coal has probably had more to do with rapid increases in personal income and falling unemployment rates in some of the hardest-hit parts of Appalachia than have SOC investments.

Efforts of the Economic Development Administration (EDA, the successor to the Area Redevelopment Administration) have been more diffuse, both geographically and in terms of programs. EDA has relied on a complex mixture of investment in directly productive activities (business loans) and public facility grants. Again, it is too early to render a final judgment about the efficacy of this approach.

During the Nixon administration, the EDA was reduced to little more than a housekeeping operation. Revenue-sharing was supposed to take over the regional development functions which had been performed by EDA. To date, EDA has not fully recovered from this emasculation, although it is on the road to recovery. While its record has been a mixed one, there is evidence that EDA has contributed to the revival of many depressed communities. But what if the thesis developed in this article stands up? What if the historical association between primary production and low per capita income is reversed? Development programs based on the Clark-Fisher model might have to be abandoned or greatly modified.

What would be the appropriate response of federal agencies if presently industrialized areas are adversely affected by high and rising energy costs? The ARC and EDA programs were designed specifically to deal with areas which had not participated in the industrialization and urbanization of America. The objective was to stimulate these two processes in Appalachia and in depressed areas elsewhere. But this would scarcely be an appropriate policy objective for regions which are not deficient in social overhead capital and whose basic problem might well become that of excess industrial capacity.

One policy alternative would be to shift the emphasis of energy research and development from low-technology areas (coal liquefaction and gasification) to high-technology subjects (the breeder reactor, fusion, and solar energy).[23] Breeder reactors and solar energy storage facilities could be located anywhere; they could be sited in regions that are presently heavily industrialized and thus guarantee a continuing supply of energy at prices that would be competitive *on a regional basis*. The problem is that even the most optimistic advocates of high-technology approaches in the search for new energy sources do not see much chance of their successful development during the present century.[24] And the nation's industrial map could be completely altered in a quarter of a century.

Energy specialists are aware of the technical difficulties yet to be overcome before such dramatic and safe sources of energy as fusion and solar energy are practical realities.[25] Thus, the Energy Research and Development Administration (ERDA) behaved realistically when it shifted emphasis from high to low technology by stepping up such programs as coal liquefaction and coal gasification and extending the time horizon on the breeder reactor.[26]

What is the answer to the problems of industrial dislocation that could be engendered by continuing rises in energy prices? There probably is none, although panaceas abound if one wishes to look for them. Technological optimists tell us not to worry. But their optimistic projections are not supported by either empirical data or logic. Consumer advocates such as Ralph Nader would have us believe that the problem is entirely the result of corporate misbehavior. If cartels could be destroyed, they argue, everything would be fine again. But those who think the world is made up entirely of bad buys and consumers rarely bother to look at supply constraints.

The major problem of adjustment may be social-psychological

rather than economic. Most of us have taken economic growth for granted.[27] Cyclical interruptions to growth, we believe, are nothing more than a manifestation of our inability to properly manage a growth economy.

A few economists—such as Georgescu-Roegen and Daley in the United States and Schumacher in Great Britain—have raised very difficult questions about the economics of growth.[28] But these questions have been largely ignored. We may be rapidly approaching the time, however, when the issues of energy supply and cost will be of central concern to economists. Energy and other resource scarcities may force us, however reluctantly, to recognize that the answer to all economic problems is not simply more growth.

To return to the regional issue, which is the basic focus of this article, neither conventional regional development policies nor reliance on market forces will permit the nation to cope with the difficulties that will be faced by energy-poor, highly industrialized regions. New policies will have to be devised to help these regions adjust to the structural changes that further increases in energy prices are almost certain to engender.

NOTES TO CHAPTER 5

1. Colin Clark, *The Conditions of Economic Progress* (London: Macmillan, 1940); Allen G.B. Fisher, *The Clash of Progress and Security* (London: Macmillan, 1935).

2. Lewis H. Bean, "International Industrialization and Per Capita Income," in part 5 of Conference on Research in Income and Wealth, *Studies in Income and Wealth* (Washington, D.C.: National Bureau of Economic Research, 1946).

3. Charles D. Hyson and Alfred C. Neal, "New England's Economic Prospects," *Harvard Business Review* 26 (March 1948): 156–80.

4. Seymour E. Harris, *The Economics of New England* (Cambridge: Harvard U. Press, 1952).

5. Jean Fourastié, *Esquisse d'une théorie générale de l'évolution économique contemporaine* (Paris: Presses Universitaires de France, 1947); *Le grand Espoir du XXe siècle* (Paris: Presses Universitaires de France, 1950).

6. An important component of the early TVA program was a plan to provide an abundance of cheap energy both for revitalizing agriculture and for stimulating the growth of an industrial sector.

7. This legislation includes the Area Redevelopment Act (1961); its successor, the Economic Development Act (1965); and the Appalachian Regional Development Act (1965).

8. U.S. Congress, Joint Economic Committee, *The Current Fiscal Position of State and Local Governments*, 94th Cong., 1st sess., 17 Decenber 1975, p. 3.

9. Ibid.

10. Energy states, as defined by the JEC, include Oklahoma, Texas, Louisi-

ana, West Virginia, Ohio, Utah, Indiana, New Mexico, Alabama, Arkansas, Montana, Wyoming, and Tennessee.

11. These states are Iowa, Minnesota, North Dakota, South Dakota, Wisconsin, Kansas, Nebraska, and Idaho.

12. JEC, op. cit., p. 4.

13. In Wyoming, West Virginia, and Louisiana, energy employment accounts for between one-fifth and one-third of export-base employment. William H. Miernyk, "Regional Employment Impacts of Rising Energy Prices," *Labor Law Journal* 26 (August 1975); 518–23.

14. The small energy producers are Alaska, North Dakota, Montana, and Nevada.

15. The exceptions are Kansas, Colorado, Virginia, Pennsylvania, California, and Illinois. See William H. Miernyk, "The Regional Economic Consequences of High Energy Prices," *Journal of Energy and Development* 1 (Spring 1976); 213–39.

16. William H. Miernyk, "The Northeast Isn't What It Used to Be," in *Balanced Growth for the Northeast*, Proceedings of a Conference of Legislative Leaders on the Future of the Northeast (Albany, New York State Senate, 1975), pp. 19–48; Miernyk, "The Regional Economic Consequences of High Energy Prices."

17. Committee for Economic Development, *Achieving Energy Independence* (New York: Committee for Economic Development, 1974).

18. For an elaboration of these views and specific references to those advancing them, see Miernyk, "The Regional Economic Consequences of High Energy Prices."

19. William H. Miernyk, "Coal and the Future of the Appalachian Economy," *Appalachia* 9 (October–November 1975): 29–35.

20. Although unions have been far more successful in some states than in others, for example, they have had relatively little impact on geographic wage and income differentials. On this, see Lloyd G. Reynolds and Cynthia H. Taft, *The Evolution of Wage Structure* (New Haven, Yale U. Press, 1956).

21. The numbers in parentheses are the 1967 input-output sector numbers. See U.S. Department of Commerce, Bureau of Economic Analysis, *Input-Output Structure of the U.S. Economy: 1967*, 3 vols. (Washington, D.C.: Government Printing Office, 1974).

22. Likely candidates are relatively footloose establishments which provide inputs to the sectors marked by an asterisk in Table 5–1.

23. H.A. Bethe, "The Necessity of Fission Power," *Scientific American* 234 (January 1976): 21–31; Paul P. Craig, "The Liquid-Metal Fast Breeder Reactor," *Energy Systems and Policy* 1 (1975): 203–204; Hans L. Hamester, Glen A. Graves, and James L. Plummer, "Risk Aversion and Energy Policy: A Case for Breeder Research and Development," *Energy Systems and Policy* 1 (1975): 233–58; William D. Nordhaus, "Resources as a Constraint on Growth," *American Economic Review* 64 (May 1974): 22–26; Robert M. Solow, "The Economics of Resources or the Resources of Economics," *American Economic Review* 64 (May 1974): 1–14.

24. David J. Rose, "Commentary on the Foregoing Breeder Papers and on the Problem in General," *Energy Systems and Policy* 1 (1975): 271–76; John P. Holdren, "Uranium Availability and the Breeder Decision," *Energy Systems and Policy* 1 (1975): 205–32; W.I. Finch et al., "Discussion of Uranium Availability and the Breeder Decision," *Energy Systems and Policy* 1 (1975): 259–70.

25. Dixy Lee Ray, *The Nation's Energy Future* (Washington, D.C.: Government Printing Office, 1973); U.S. Congress, Office of Technology Assessment, Committees on Science and Technology, Senate Interior and Insular Affairs, and the Joint Economic Committee on Atomic Energy, *An Analysis Identifying Issues in the Fiscal Year 1976 ERDA Budget*, Joint Economic Print Serial D (Washington, D.C.: Government Printing Office, 1975).

26. For an example of the continued confusion surrrounding the breeder reactor, see Mark Sharefkin, *The Fast Breeder Reactor Decision: An Analysis of Limits and the Limits of Analysis*, Joint Economic Committee (Washington, D.C.: Government Printing Office, 1976).

27. I am aware, of course, that there have been scattered attacks on the adverse environmental consequences of economic growth by Mishan and others, but those who have been critical of growth have been a very small minority of the economics profession.

28. Herman E. Daly, "The Economics of the Steady State," *American Economic Review* 64 (May 1974): 15–21; Nicholas Georgescu-Roegen, "Energy and Economic Myths," *Southern Economic Journal* 41 (January 1975): 347–81; Ernst F. Schumacher, *Small Is Beautiful: Economics as if People Mattered* (New York: Harper & Row, 1973). See also Dennis L. Meadows et al., *Dynamics of Growth in a Finite World* (Cambridge, Mass.: Wright-Allen, 1974).

Chapter 6

Application of an Interindustry Supply Model to Energy Issues

Frank Giarratani

INTRODUCTION

Input-output applications to energy issues have, up until now, concentrated on demand relationships. The system of linear relations that is used to represent the structure of an economy is solved for its reduced form and relates the total (gross) output requirement of each sector to its final (net) use in all sectors. As the vector of final output changes, interindustry flows are allowed to change subject only to technical factors of production. Within this framework, Kutscher and Bowman (1974) have examined the direct and total dollar energy requirements of the United States, and related these flows to sectoral employment in order to analyze the potential impact of petroleum shortages on the labor force. Herendeen (1973) converted the detailed 1963 U.S. tables to Btu equivalents and calculated the total energy cost of goods and services. Other energy accounting studies have been undertaken in Great Britain (Wright, 1975) and Canada (Gupta and Capel, 1975).

Studies of this nature by no means exhaust the possibilities of energy-related interindustry analysis. Ghosh (1958) demonstrated that a model may be formulated to relate total output to *supply* factors. Augustinovics (1970) has generalized this framework in a rigorous comparison of the demand and supply systems. As will be shown below, an analysis of interindustry flows, from the supply perspective, can yield important insights into energy-related issues.

Reprinted by permission from *Environment and Planning A*, Volume 8, No. 4 (June 1976): 447–54.

The input-output table is a neutral image of an economy, emphasizing neither supply nor demand forces but rather recording equilibrium values at one point in time. Quite simply, it is a social accounting array, with details of industrial transaction, based on identities that equate the value of each sector's output to the value of its inputs; thus

$$x = \mathbf{X}i + y, \tag{1}$$

and

$$x = \mathbf{X}'i + v, \tag{2}$$

where

$\mathbf{X} = [x_{ij}]$, is an input-output table of intermediate flows from sector i to sector j, and $\mathbf{X}'$ is its transpose,

$y = [y_i]$, is a vector of final use,

$v = [v_i]$, is a vector of primary inputs (value added),

$x = [x_i]$, is a vector of gross output, and

i is the unit vector, for $i, j = 1, \ldots, n$ industrial sectors.

AN INTERINDUSTRY MODEL

As usually formulated, the input-output *table* is transformed into an analytical *model* by substituting the linear-demand relationship,

$$x_{ij} = a_{ij}x_j, \qquad i, j = 1, \ldots, n, \tag{3}$$

into the identity (1) and solving for gross output as a function of final use. The coefficients a_{ij} are the familiar Leontief production or technical coefficients. In accordance with Ghosh (1958, p. 58), we may also assume, for all i, j that

$$x_{ij} = \vec{a}_{ij}x_i. \tag{4}$$

In this case, interindustry flows are assumed to be proportional to the output of the producing (selling) industries. This model implies that production relationships are determined by the availability of inputs rather than by technical factors.

The behavioral assumptions in equations (3) and (4) are, of course, subject to criticism. It is widely recognized that factor substitution is possible and that the fixed production relationship implicit in equation (3) very probably holds only in the short run. The usefulness of this assumption clearly rests on the existence of unused capacity and very elastic factor-supply curves. However, the ability of productive processes to substitute commodity factors to the degree implied by equation (4) is no less questionable. Ghosh (1958, p. 59) has argued that under monopolistic market conditions in which resources are scarce, the allocation function, defined by equation (4), will be dominant. Empirical evidence presented by Augustinovics (1970, p. 260–63) seems to refute the belief that input coefficients, a_{ij}, are more stable than output coefficients, $\vec{a}_{ij}$. Temporal changes in these coefficients for Hungarian data are of the same magnitude.

Taking the transpose of identity (2)—for convenience of computation—and converting equation (4) to matrix notation, we have

$$x' = i'X + v', \qquad (2')$$

and

$$i'X = x'\vec{A}, \qquad (5)$$

where $\vec{A} = [\vec{a}_{ij}]$ are the output coefficients which indicate direct sales from sector i to sector j per unit output of sector i. Given equation (5), the supply identity may be solved for gross output in terms of primary inputs and structural parameters:

$$x' = v'(I - \vec{A})^{-1}, \qquad (6)$$

or

$$x' = v'\vec{Q}, \qquad (7)$$

where $\vec{Q} = (I - \vec{A})^{-1}$. From equation (7) we may take

$$\frac{\partial x_j}{\partial v_i} = \vec{Q}_{ij}. \qquad (8)$$

The elements $\vec{Q}_{ij}$ are supply multipliers relating unit changes in primary inputs to changes in gross output (see Hoover, 1971, pp. 235–37).

An intuitive explanation of these multipliers parallels the more familiar demand relationship. Changes in the level of primary factors available to a given sector permit changes in its level of output, as well as in the levels of output of sectors requiring its product for intermediate use. This first-round effect and all subsequent rounds of industrial activity are embodied in the supply multiplier $\vec{Q}_{ij}$. Given autonomous inputs, this model assumes that demand is always forthcoming, and is the inverse of the conventional input-output model.

As will be shown below, energy-related interindustry analysis that emphasizes supply relationships can proceed on several levels. The hierarchy is determined by the degree of complexity of the coefficients analyzed. Consider the simple case of the matrix $\vec{Q}$. Each element $\vec{Q}_{ij}$ measures the output in sector j that is permitted directly and indirectly by a unit increase in primary inputs of sector i. Row sums of the matrix $\vec{Q} = \Sigma_j \vec{Q}_{ij}$ are multipliers describing the total output permitted in the *economy* by that unit increase.

APPLICATION OF THE MODEL
TO U.S. DATA

The U.S. interindustry table for 1967 (BEA, 1974) includes two extractive energy activities, Coal mining, and Crude petroleum and natural gas, and one fabricating activity, Petroleum refining and related industries. An examination of the matrix of direct and indirect effects, $\vec{Q}$, for the United States shows the mutually dependent relationship of the energy sectors and the economy as a whole. When total supply multipliers (row sums of $\vec{Q}$) are calculated and ranked by size (see Table 6–1), the crucial position of the extractive energy sectors is demonstrated. These sectors rank ninth and eleventh in seventy-eight national sectors.[1] It is well recognized that supply constraints in energy sectors can severely limit potential national output. As indicated by these multipliers, shortages of primary factors in extractive energy sectors can have approximately a threefold effect on the national economy.

Within this framework it is also possible to identify supply linkages that have the potential of significantly limiting *energy* output. From equation (7) we wish to examine equation (8),

$$\frac{\partial x_j}{\partial v_i} = \vec{Q}_{ij},$$

for j = 7, 8, and 31 (the energy sectors) over all sectors i (see Table 6–2). Interindustry supply constraints are clearly more important for the petroleum-refining sector than for the extractive energy activities. Coal, for example, has particularly weak interindustry supply

Table 6-1.   Total Supply Multipliers of Twenty Top-Ranked U.S. Sectors[a]

Sector Number	Sector Name	Total Supply Multipliers	Rank
5	Iron and ferroalloy ores mining	4.0106	1
6	Nonferrous metal ores mining	3.9546	2
67	Radio and TV broadcasting	3.5115	3
10	Chemical and fertilizer mineral mining	3.4993	4
28	Plastic and synthetic materials	3.2469	5
38	Primary nonferrous metal manufacturing	3.2116	6
4	Agriculture, forestry and fishery services	3.1879	7
27	Chemical and chemical products	3.1438	8
7	Coal mining	3.1285	9
37	Primary iron and steel manufacturing	3.1158	10
8	Crude petroleum and natural gas	3.0927	11
9	Stone and clay mining, and quarrying	3.0496	12
24	Paper and allied products, except containers	2.9487	13
3	Forestry and fishery products	2.9433	14
30	Paints and allied products	2.9382	15
16	Broad and narrow fabrics	2.9080	16
21	Wooden containers	2.8502	17
50	Machine shop products	2.8491	18
25	Paperboard containers and boxes	2.8068	19
20	Lumber and wood products, except containers	2.8012	20

[a]Range of total supply multipliers: the largest is Iron and ferroalloy ores mining, value 4.0106; the smallest is Medical, educational services and nonprofit organizations, value 1.0922.

Table 6–2.   Selected Energy-Sector Supply Multipliers[a]

Sector	Coal Mining $\vec{Q}_{i,7}$	Crude Petroleum and Natural Gas $\vec{Q}_{i,8}$	Petroleum Refining and Related Industries $\vec{Q}_{i,8}$	Total
5 Iron and ferroalloy ores mining	0.0039	0.0096	0.0146	0.0282
6 Nonferrous metal ores mining	0.0019	0.0053	0.0126	0.0198
7 Coal mining	1.1470	0.0066	0.0202	1.1740
8 Crude petroleum and natural gas	0.0025	1.0311	0.8605	1.8942
9 Stone and clay mining, and quarrying	0.0029	0.0092	0.0459	0.0582
10 Chemical and fertilizer mineral mining	0.0029	0.0090	0.0353	0.0473
12 Maintenance and repair construction	0.0025	0.0305	0.0504	0.0835
24 Paper and allied products, except containers	0.0008	0.0029	0.0171	0.0210
25 Paperboard containers and boxes	0.0008	0.0024	0.0203	0.0235
26 Printing and publishing	0.0013	0.0046	0.0146	0.0206
27 Chemical and chemical products	0.0040	0.0122	0.0513	0.0677
28 Plastic and synthetic materials	0.0018	0.0039	0.0118	0.0176
30 Paints and allied products	0.0014	0.0144	0.0297	0.0456
31 Petroleum refining and related industries	0.0023	0.0049	1.0847	1.0921
32 Rubber and miscellaneous plastic products	0.0030	0.0051	0.0132	0.0215
36 Stone and clay products	0.0019	0.0123	0.0202	0.0346
37 Primary iron and steel manufacturing	0.0040	0.0100	0.0142	0.0283
38 Primary nonferrous metal manufacturing	0.0019	0.0050	0.0115	0.0185
39 Metal containers	0.0007	0.0026	0.0369	0.0402
42 Other fabricated metal products	0.0021	0.0089	0.0133	0.0244
43 Engines and turbines	0.0056	0.0165	0.0162	0.0383
45 Construction, mining, and oil-field machinery	0.0210	0.0124	0.0128	0.0462
46 Materials handling machinery and equipment	0.0010	0.0031	0.0135	0.0177
49 General industrial machinery and equipment	0.0025	0.0154	0.0257	0.0437
50 Machine shop products	0.0039	0.0261	0.0254	0.0555

53	Electrical industrial equipment and apparatus	0.0007	0.0206	0.0195	0.0409
55	Electrical lighting and wiring equipment	0.0028	0.0047	0.0101	0.0178
65	Transportation and warehousing	0.0013	0.0056	0.0392	0.0462
67	Radio and TV broadcasting	0.0025	0.0086	0.0274	0.0386
68	Electricity, gas, water and sanitary services	0.0033	0.0082	0.0272	0.0388
70	Finance and insurance	0.0014	0.0055	0.0142	0.0212
71	Real estate and rental	0.0020	0.0239	0.0283	0.0543
73	Business services	0.0026	0.0086	0.0276	0.0388
77	Federal government enterprises	0.0014	0.0052	0.0137	0.0204
78	State and local government enterprises	0.0022	0.0077	0.0235	0.0334

[a]Includes all sectors with at least one energy-supply multiplier in excess of 0.0100 (all numbers have been rounded from six decimal places).

linkages. Only one related activity (sector 45) exhibits a supply multiplier of any consequence. As with crude petroleum, significant linkages are few in number and small in size. This is clearly not the case in the petroleum-refining sector. Several supplying industries have multipliers in excess of 0.04 for unit changes in primary inputs.

An array of multipliers, similar to those in Table 6-2, could also be calculated by relating value-added changes in supplying sectors to *final* energy use. In the supply model, final use may be calculated directly, given equilibrium values of gross output (just as employment is calculated in the traditional demand model) since, by assumption, they are related linearly:

$$\langle y \rangle = \langle \vec{y} \rangle \langle x \rangle, \tag{9}$$

where the notation $\langle \rangle$ denotes a diagonal matrix for the associated vector. Final-use supply multipliers may be taken from the elements of

$$\vec{Q} \langle \vec{y} \rangle = [\, \vec{Q}_{ij} \vec{y}_j \,] \,. \tag{10}$$

Each element represents total *final* use in sector j permitted per unit value added in sector i. These complex coefficients would reflect the dominance of the petroleum-refining sector, with respect to final energy use, by weighting its supplying sectors proportionately more than those of the extractive energy sectors.

This model can also be applied to the problem of allocating scarce energy resources to intermediate users. In a period of extreme shortage, it becomes relevant to question the validity of treating the production of scarce resources endogenously. If, as in the present case, we are concerned with energy production and its impact on the economy, we might hold petroleum-related sectors (Crude petroleum and natural gas, and Petroleum refining and related industries) exogenous by deleting these activities from the interindustry matrix, $\vec{A}$.

The calculations carried out for equation (6) may be repeated where the vector of primary inputs has been augmented to include exogenous energy inputs, and where its length, as well as that of the gross output vector, has been correspondingly reduced; thus

$$x' = v'_1 \, (I - \vec{A})^{-1} \tag{11}$$

$$= v'_1 \, \vec{Q} \,, \tag{12}$$

where

$$v'_1 = v' + x'_e \,, \tag{13}$$

and $x_e = [x_{ei}]$ is a vector of exogenous energy inputs to industrial sectors (for all $i, j = 1, \ldots, n - 2$).[2]

The partial derivatives $\partial x_j / \partial v_{1i}$ have the same meaning as above, but it is now possible to concentrate on primary energy inputs, x_{ei}.

For example, the impact of alternative allocation programs can be examined. Consider the case where petroleum supplies are to be distributed on the basis of the existing pattern of sales; that is, each purchasing sector is to receive a share equivalent to its proportion of total petroleum use before the shortage existed; thus

$$\vec{a}_{ei} = \frac{x_{ei}}{\sum_i x_{ei} + y_e} , \tag{14}$$

where y_e is the final petroleum use.

The model suggests that a dollar of petroleum production distributed in this fashion will lead to a total output in each industrial sector of

$$x' = \vec{a}'_e \vec{Q} , \tag{15}$$

where $\vec{a}'_e$ is a vector composed of elements $\vec{a}_{ei}$.

The vector x' of equation (15) is simply the column sum of the matrix

$$\langle \vec{a}_e \rangle \vec{Q} = [\vec{a}_{ei} \vec{Q}_{ij}] . \tag{16}$$

Each element of this matrix represents the output in sector j permitted if the ith supplying sector receives its "share" of a unit of petroleum supplies. As demonstrated earlier, column sums of this matrix (summation over i) offer information on the total output in each sector, j, forthcoming from a unit distribution of petroleum under this allocation plan. Row sums of this matrix (summation over j) offer information on the total output permitted in the economy if sector i receives its "share" of the unit to be distributed. Both sets of information may be valuable in choosing among alternative distribution plans.

The results of these calculations are presented in Table 6-3. As indicated, one dollar of petroleum supplies delivered under this allocation structure allows total production in excess of 1.37. Alternative petroleum-distribution plans—as between intermediate and final use, and among producing activities—would obviously have different total effects. The potential impact of a number of plans could be effectively compared using simulations of this nature.

Table 6—3.   Supply Model Multipliers with Exogenous Petroleum Sector

Sector	Direct Primary Energy Input Allocation	Total Sector Output Per Unit Petroleum Distribution (column sums)	Total Economy Output Per Unit Petroleum Distribution (row sums)
1 Livestock and livestock products	0.0040	0.0224	0.0104
2 Other agricultural products	0.0216	0.0336	0.0548
3 Forestry and fishery products	0.0009	0.0012	0.0025
4 Agricultural, forestry, and fishery services	0.0001	0.0011	0.0004
5 Iron and ferroalloy ores mining	0.0001	0.0006	0.0004
6 Nonferrous metal ores mining	0.0001	0.0006	0.0005
7 Coal mining	0.0009	0.0016	0.0026
9 Stone and clay mining, and quarrying	0.0018	0.0024	0.0052
10 Chemical and fertilizer mineral mining	0.0001	0.0003	0.0002
11 New construction	0.0333	0.0541	0.0333
12 Maintenance and repair construction	0.0148	0.0194	0.0352
13 Ordnance and accessories	0.0006	0.0031	0.0008
14 Food and kindred products	0.0052	0.0474	0.0073
15 Tobacco manufactures	*	0.0028	*
16 Broad and narrow fabrics	0.0005	0.0098	0.0015
17 Miscellaneous textile goods and floor covering	0.0001	0.0027	0.0003
18 Apparel	0.0005	0.0079	0.0007
19 Miscellaneous fabricated textile products	*	0.0018	0.0001
20 Lumber and wood products, except containers	0.0025	0.0072	0.0070
21 Wooden containers	*	0.0002	0.0001
22 Household furniture	0.0002	0.0017	0.0002
23 Other furniture and fixtures	0.0001	0.0008	0.0001
24 Paper and allied products, except containers	0.0024	0.0093	0.0071
25 Paperboard containers and boxes	0.0007	0.0032	0.0019
26 Printing and publishing	0.0010	0.0065	0.0024
27 Chemical and chemical products	0.0440	0.0612	0.1333
28 Plastic and synthetic materials	0.0028	0.0122	0.0091
29 Drugs, cleaning, and toilet preparations	0.0013	0.0074	0.0019

No.	Industry	Col 1	Col 2	Col 3
30	Paints and allied products	0.0031	0.0036	0.0011
32	Rubber and miscellaneous plastic products	0.0012	0.0075	0.0005
33	Leather tanning and leather products	0.0002	0.0006	0.0001
34	Footwear and other leather products	0.0001	0.0012	0.0001
35	Glass and glass products	0.0004	0.0014	0.0002
36	Stone and clay products	0.0059	0.0072	0.0025
37	Primary iron and steel manufacturing	0.0085	0.0128	0.0028
38	Primary nonferrous metal manufacturing	0.0037	0.0073	0.0011
39	Metal containers	0.0002	0.0011	0.0001
40	Heating, plumbing, and metal products	0.0017	0.0040	0.0008
41	Stampings, screw machine products	0.0011	0.0026	0.0004
42	Other fabricated metal products	0.0026	0.0042	0.0010
43	Engines and turbines	0.0007	0.0012	0.0004
44	Farm machinery and equipment	0.0004	0.0013	0.0003
45	Construction, mining, and oil-field machinery	0.0008	0.0017	0.0005
46	Materials handling machinery and equipment	0.0002	0.0007	0.0001
47	Metalworking machinery and equipment	0.0020	0.0025	0.0010
48	Specialized industry machinery and equipment	0.0010	0.0017	0.0006
49	General industrial machinery and equipment	0.0015	0.0023	0.0007
50	Machine shop products	0.0029	0.0018	0.0010
51	Office, computing, and accounting machinery	0.0002	0.0012	0.0002
52	Service industry machines	0.0005	0.0016	0.0003
53	Electrical industrial equipment and apparatus	0.0021	0.0031	0.0011
54	Household appliances	0.0003	0.0017	0.0002
55	Electrical lighting and wiring equipment	0.0003	0.0011	0.0001
56	Radio, TV, and communication equipment	0.0008	0.0034	0.0006
57	Electronic components and accessories	0.0006	0.0021	0.0003
58	Miscellaneous electrical machinery and equipment	0.0002	0.0009	0.0001
59	Motor vehicles and equipment	0.0027	0.0126	0.0017
60	Aircrafts and parts	0.0032	0.0061	0.0021
61	Other transportation equipment	0.0004	0.0022	0.0003
62	Scientific and controlling instruments	0.0005	0.0015	0.0003
63	Optical, ophthalmic and photographic equipment	0.0002	0.0013	0.0001
64	Miscellaneous manufacturing	0.0011	0.0031	0.0007

(Table 6–3 continued overleaf)

Table 6–3. continued

Sector	Direct Primary Energy Input Allocation	Total Sector Output Per Unit Petroleum Distribution (column sums)	Total Economy Output Per Unit Petroleum Distribution (row sums)
65 Transportation and warehousing	0.0482	0.0601	0.0956
66 Communication, except radio and TV	0.0015	0.0030	0.0029
67 Radio and TV broadcasting	*	0.0004	*
68 Electricity, gas, water, and sanitary services	0.0666	0.0896	0.1439
69 Wholesale and retail trade	0.0327	0.0526	0.0478
70 Finance and insurance	0.0022	0.0074	0.0039
71 Real estate and rental	0.0210	0.0374	0.0330
72 Hotels, personal and repair services	0.0055	0.0094	0.0070
73 Business services	0.0031	0.0115	0.0076
74 Automobile repair and services	0.0041	0.0066	0.0072
75 Amusement	0.0004	0.0021	0.0007
76 Medical, educational services and nonprofit organizations	0.0050	0.0134	0.0055
77 Federal government enterprises	0.0011	0.0026	0.0025
78 State and local government enterprises	0.0023	0.0072	0.0062
Final use	0.6433	0.6433	0.6433
Total	1.0000	1.3777	1.3777

*Less than 0.0001 (all numbers have been rounded from six decimal places).

CONCLUSION

These empirical results are meant to demonstrate the potential of this aspect of interindustry analysis to energy issues. It is not suggested that the model represents a definitive framework for related analysis, but simply that it has the potential of contributing significantly to our understanding of some aspects of the issues addressed. This point should be emphasized with respect to the problem of comparing energy-allocation alternatives. Consideration of the impact of various strategies on the final use of all commodities is particularly neglected by this model. The tables summarize a large number of calculations that could be productively analyzed in a number of ways. When the model is applied to existing interindustry tables for energy-producing and energy-consuming regions, insights may yet be gained at that important level of analysis.

NOTES TO CHAPTER 6

William H. Miernyk, M.J. Hwang, and Charles Socher very kindly offered their criticism on an earlier draft. Mr. Socher programmed and carried out the computations described in this paper. For their assistance I owe my sincere thanks. This work was supported by the Economic Development Administration, U.S. Department of Commerce, under Grant OER 519−G−76−9.

1. The national model actually includes eighty-two endogenous sectors. For this paper, all dummy sectors have been omitted. It should also be pointed out that an accounting convention of the national model allocates imported goods directly to competing U.S. sectors. It is therefore not possible to distinguish between domestic and imported energy flows.

2. Because of aggregation in the national model, it is not possible to create separate sectors for natural gas and crude petroleum. The dimension of all relevant vectors and matrixes has been reduced by a factor of two, the number of newly defined exogenous sectors.

REFERENCES

Augustinovics, M. 1970. "Methods of International and Intertemporal Comparison of Structure." In *Contributions to Input-Output Analysis*, eds. A.P. Carter, and A. Brody. Amsterdam: North-Holland, pp. 249−69.

BEA. 1974. "The Input-Output Structure of the U.S. Economy, 1967." *Survey of Current Business* 54 (2): 24−56 (Bureau of Economic Analysis, U.S. Department of Commerce).

Ghosh, A. 1958. "Input-Output Approach to an Allocative System." *Econometrica* 25 (97): 58−64.

Gupta, R.P., Capel, R.E. 1975. "An Inter-Industry Analysis of National and Interregional Energy Allocation in Canada: A Proposal and Preliminary Results."

Paper presented at the Mid-Continent Section, Regional Science Association Annual Meeting, Duluth, Minnesota.

Herendeen, R.A. 1973. *The Energy Cost of Goods and Services.* ORNL–NSF–EP–58 (Oak Ridge National Laboratory, Oak Ridge, Tenn.).

Hoover, E.M. 1971. *An Introduction to Regional Economics.* New York: Alfred A. Knopf.

Kutscher, R.E., Bowman, C.T. 1974. "Industrial Use of Petroleum: Effect on Employment. *Monthly Labor Review* 97 (3): 3–8.

Wright, D.J. 1975. "The Natural Resource Requirements of Commodities." *Applied Economics* 7 (1): 31–39.

Chapter 7

The Pattern of Industrial Location and Rising Energy Prices

Frank Giarratani
Charles F. Socher

High and rising energy prices are likely to affect the national pattern of consumption and production to a significant extent. The impact of energy-induced price increases on individual consumer demand is already evident; however, the effect of these changes on the pattern of industrial location is likely to be more subtle. The focus of this note is on anticipated changes in the distribution of industry between energy-producing and energy-consuming regions.

Chinitz has identified two historical phases in the distribution of production that are shaped largely by structural changes in transportation [6]. The nineteenth century was a period of increased concentration of many manufacturing industries. During this period, the declining cost and increased speed of freight transport, due primarily to the advent of a national rail network, had a substantial impact on the relative cost of long- versus short-haul transport. Major economies in long-distance freight shipments made it possible to move away from local markets to raw material sites and exploit the technological advantages of mass production.

The second phase discussed by Chinitz is a period of decentralization of industry beginning early in the twentieth century. During this period, major shifts in the regional distribution of manufacturing activity are attributed largely to transport factors. He argues that a

Reprinted by permission from *Atlantic Economic Journal*, Vol. V, No. 1 (March 1977).

This research was supported by the Economic Development Administration, U.S. Department of Commerce, under Grant No. OER–519–G–76–9 (99–7–13327).

general increase in the cost of transportation *relative* to the cost of other factors of production encouraged a decentralization of industry and favored locations closer to the market. Technological developments that reduced the cost of short hauls relative to long hauls—particularly the advent of truck transport—supported this trend.[1]

CLASSICAL LOCATION ANALYSIS

The foundation for much of the substantive analysis on transport costs as a determinant of industrial location was laid by Weber [20]. In his model, price and quantity demanded of the product are fixed at the market [20, pp. 37–39]. The location of raw materials is given; and labor is assumed to be immobile, with fixed wages and its supply at each location unlimited. With these assumptions and relying on fixed factor proportions in production, profits are maximized at the least transport cost location.

Hoover extended Weber's work to include realistic rate structures [11, 13]. His discussion of the relationship between transport costs and distance may be explained by distinguishing two general components of transfer costs: terminal and line-haul costs. Terminal costs are invariant with distance and include costs such as loading and unloading, packaging, and associated paperwork. Line-haul costs are those directly related to movement over space and therefore can be expected to vary directly, though less than proportionally, with distance. Hoover [13, p. 19] points out that empirically the

> . . . tendency of transfer costs to "taper off" with increasing distance is characteristic of all agencies, but naturally much more marked in those which need a heavy investment in terminal facilities. . . . [See Figure 7–1.]

The choice of an appropriate transport medium or combination of media will depend, in part, on the distance the goods are to be shipped so as to take advantage of potential long-haul economies. This has been indicated in Figure 7–1 by the heavy line showing the effective choice for various lengths of haul.

Each of the major transport media—motor carrier, railroad, and ship transport—will be affected to some extent by fuel price increases. Data presented in Table 7–1 show their respective Btu requirements per ton-mile. Motor carrier transport is relatively energy inefficient, using approximately four times the energy per ton-mile as railroads or water transport. Returning briefly to Figure 7–1, absolute increases in transport costs for all agencies accompanied by a relative increase in line-haul costs of motor-carrier relative to rail

Figure 7-1. Transport Costs and Medium of Transport

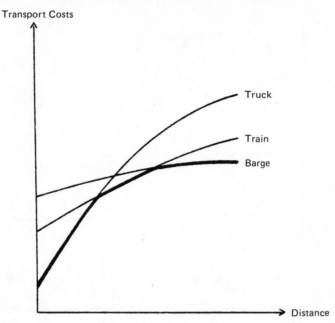

Table 7-1. Energy Efficiency for Selected Intercity Freight Transport

	Ton-Mile Per Gallon	Btu Per Ton-Mile[a]
Water	250	540
Rail	200	680
Motor	58	2,340

Source: Hirst [10].
[a] Assuming 136,000 Btu/gallon.

and ship transport would cause an upward shift and increased convexity in the effective curve of transport costs.

As demonstrated by Hoover, the effect of a rate structure that increases less than proportionally with distance is to encourage plant location at either raw material sites or the market, rather than at some intermediate point [13, p. 31]. As an illustration, consider the simple case of an activity that uses only one material and sells its product at a single market (Figure 7-2). Total transport costs consist of procurement costs (getting the material to the place of production) and distribution costs (sending the product to the market).

Figure 7−2. Total Transport Costs

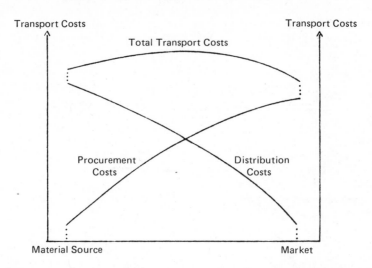

Market or material orientation is then determined by differences between transport costs on the commodities shipped. Location is at the point where total transport costs are minimized (the market in Figure 7−2). Economies of long hauls imply that intermediate points of production are likely to incur higher total transport costs. Consideration of terminal charges, represented by dotted lines in Figure 7−2, also encourages an end-point solution since location at either extreme would save one set of terminal costs.

Under the assumption that multiple transport media are used by firms, this model would therefore argue for increased dominance of end-point locations. Beyond this, however, changes in relative freight rates on material and product can have a significant effect on *material* versus *market* orientation. Hoover cites the American flour-mill industry as a case in point [11, p. 38]. More recently, electric power production in the United States has undergone a rather pronounced shift toward raw material locations primarily because of reductions in the cost (and efficiency) of transmitting electric energy.

This kind of analysis underlies much of the work by Chinitz and others who have argued that historically the market has become an increasingly dominant factor in explaining the spatial distribution of activity (see note 1). Similarly, to the extent that rising energy prices affect the level of transport costs or have differential effects on transport agencies predominantly associated with one class of commodities (e.g., raw materials or consumption goods), the spatial distribution of production may be altered.

Raw materials and finished goods are primarily dependent on quite different transport agencies. Low value to weight and bulky materials are, for the most part, transported by rail or ship, while high value to weight—typically more processed goods—are more dependent on truck transport.[2] Pressure on transport rates due to energy price increases can therefore be expected to be substantially higher on more processed goods. Their relative dependence on truck transport is an important factor. Also, producers of high value to weight goods are less sensitive to transport rate increases. A smaller portion of costs per dollar of output is related to transport inputs; hence, changes in transport rates have limited effects on total costs. These cost changes can easily be passed on to users.

Given the Weberian model of industrial location, the implication of this structural change in transport rates is well defined. The "pull" associated with each material input site, as well as the market's "pull," is described in terms of ideal weights: coefficients of the fixed proportion production function weighted by associated transport rates. The model would predict increased attraction of the market relative to material inputs—particularly raw materials—as rates for agencies transporting finished goods increase more than proportionately.

A CRITIQUE

Criticisms by Moses [15] and Alonso [1] offer perspective on these results. Relaxing the assumption of fixed coefficients in production is sufficient to obscure the Weberian analysis. As shown by Moses, the industrial location problem may be analyzed within the traditional theory of the firm. Consider the locational triangle depicted in Figure 7–3, where M_1, M_2, and C represent material sources and the market point, respectively. A locational choice within this triangle

Figure 7–3. Locational Triangle

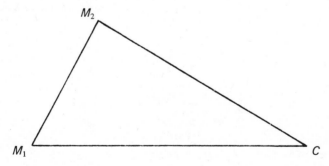

implies changes in the delivered price of inputs and, therefore, in their relative importance as a locational determinant. In Weberian terms, the ideal weight of any input is itself a function of plant location.

The explicit relationship between Weber's minimum transport cost model and the more general model admitting factor substitution has been demonstrated by Alonso [1, pp. 28–31]. Following this analysis, consider these definitions for a given firm:

P = the price of its product at the market;

Q = the quantity of the product sold;

$B(x,y)$ = delivery cost function per unit of product (a single form might be rs, where r = transport rate and s = distance from (x,y) to the market);

$p_i(x,y)$ = delivery price function of input i; e.g., $p_i(x,y) = p_i^* + r_i s_i$, where p_i^* = input i's price at its own site; and

A_i = quantity of i used.

Maximizing profits,

$$G = Q(P - B) - \sum_i p_i A_i \, , \tag{1}$$

implies examining the profit surface $G(x,y)$ for

$$\frac{\partial G}{\partial x} = \frac{\partial G}{\partial y} = 0 \, .$$

Differentiating, with respect to x,

$$\frac{\partial G}{\partial x} = -Q \frac{\partial B}{\partial x} - \sum_i A_i \frac{\partial p_i}{\partial x} - \sum_i \left[p_i \left(\sum_j \frac{\partial A_i}{\partial p_j} \cdot \frac{\partial p_j}{\partial x} \right) \right] \tag{2}$$

and rearranging terms yields

$$\frac{\partial G}{\partial x} = -Q \frac{\partial B}{\partial x} - \sum_i \left[\frac{\partial p_i}{\partial x} \left(A_i + \sum_j p_i \frac{\partial A_j}{\partial p_i} \right) \right] \, . \tag{3}$$

Substituting

$$e_{ij} = \frac{\partial A_j}{\partial p_i} \frac{p_i}{A_j},$$

we have

$$\frac{\partial G}{\partial x} = -Q \frac{\partial B}{\partial x} - \Sigma \left[\frac{\partial p_i}{\partial x} A_i (1 + \sum_j e_{ij}) \right] . \qquad (4)$$

Alonso identifies the market pull as

$$Q \frac{\partial B}{\partial x}$$

and the pull of the materials as

$$\sum_i \left[\frac{\partial p_i}{\partial x} A_i (1 + \sum_j e_{ij}) \right] .$$

The pull of materials is modified by factor substitution. To the extent that substitution in production is small (i.e., $e_{ij} \rightarrow 0$), the Weberian results remain dominant. We might make the case that with respect to material versus market orientation, substitution between raw materials and more highly processed goods is unlikely, and the pull of the materials will approach the Weberian pull,

$$\sum_i \frac{\partial p_i}{\partial x} A_i .$$

As demonstrated earlier, an increasing emphasis on market orientation is implied.

MARKET SHIFTS

In focusing on the impact of changes in the relative price of transport agencies, we should not neglect the importance of regional shifts in demand induced by energy price increases. Energy-producing states are already realizing substantial gains in personal income that may lead to further changes in the existing pattern of industrial location.

As an indication of the extent of this change, income potentials for each state have been computed for the years 1970 and 1974.

The concept of potential is borrowed from the physical sciences and was first used in location analysis by Stewart [17] and broadened by Warntz [19]. As a measure of demand, any income concentration exerts an influence that varies directly with its size and inversely with distance to other concentrations. This force field— i.e., income potential—in region i can be approximated by

$$p_i = \sum_{j=1}^{u} (Y_j/d_{ij}) \quad (i = 1, \ldots, n),$$

where

P_i = income potential in region i;

Y_j = personal income in region j; and

d_{ij} = distance from region i to region j.[3]

The percentage change in state income potential over the period 1970 to 1974 is shown in Table 7−2. The mean percentage change was 47.6 percent. For the 34 energy-producing states, the mean change was 48.4 percent, and for the 15 nonproducing states it was 45.9 percent. Thus, the data suggest that from the perspective of

Table 7−2. State Income Potential, 1970 and 1974 *(billions of dollars per mile)*

State	1970 Income Potential[a]	1974 Income Potential[a]	% Change Income Potential
Alabama*	11.2	16.9	51.13
Arizona	7.1	11.9	67.88
Arkansas**	6.5	10.1	55.12
California*	89.2	126.6	41.94
Colorado*	9.2	14.7	60.28
Connecticut	16.7	22.7	36.30
Delaware	4.8	7.0	45.28
District of Columbia	6.4	8.4	30.14
Florida*	25.7	44.6	74.02
Georgia*	16.5	24.7	49.95
Idaho	7.9	4.7	63.82
Illinois*	51.3	71.3	39.02
Indiana*	21.2	29.8	40.31
Iowa*	11.7	16.6	42.13
Kansas*	9.7	13.7	42.00
Kentucky**	11.4	16.9	48.46

Table 7-2.   continued

State	1970 Income Potential[a]	1974 Income Potential[a]	% Change Income Potential
Louisiana**	12.0	17.8	48.31
Maine	4.3	6.3	46.63
Maryland*	19.1	27.4	43.56
Massachusetts	26.2	36.0	37.37
Michigan*	38.1	55.4	45.36
Minnesota	15.6	22.5	44.10
Mississippi*	6.7	10.3	52.40
Missouri*	18.6	25.8	38.40
Montana*	2.9	4.4	50.57
Nebraska*	6.6	9.5	44.08
Nevada*	2.8	4.2	51.78
New Hampshire	4.2	6.1	45.11
New Jersey	36.6	50.8	38.83
New Mexico**	3.8	5.5	45.70
New York*	88.3	114.3	29.38
North Carolina	17.6	26.7	52.06
North Dakota*	2.5	4.5	76.31
Ohio*	43.9	61.2	39.49
Oklahoma**	9.4	13.7	44.78
Oregon*	8.3	12.6	52.70
Pennsylvania*	49.6	68.6	38.38
Rhode Island	5.8	7.9	35.66
South Carolina	8.7	13.6	55.94
South Dakota*	2.8	4.2	49.79
Tennessee*	13.2	20.3	54.52
Texas**	40.9	60.6	48.33
Utah*	4.0	6.1	51.31
Vermont	2.8	3.9	38.96
Virginia**	18.5	28.4	53.19
Washington*	14.0	20.5	45.90
West Virginia**	6.7	9.8	46.76
Wisconsin	18.0	26.2	45.23
Wyoming**	1.9	3.0	53.62

*Energy Producing States.
**Energy Surplus States.
[a]U.S. Department of Commerce, *Survey of Current Business*, August 1971, 1975; Rand McNally & Company, *Road Atlas*, 1972.

market demand, energy-producing states may be becoming relatively more attractive industrial locations. The same tendency is evident when comparing energy-surplus states to energy-deficit states. The mean increase for the former was 49.3 percent compared to the latter's 47.2 percent. These conditions exhibit a reversal of those characterizing the 1965–1970 period where the mean increase for energy-producing states was 50.3 percent versus 52.3 percent for nonproducing states.

MANUFACTURING SHIFTS

Though the changes in the patterns of industrial location may reasonably be expected to lag behind market shifts, current industrial location data offer insights to the direction, if not the magnitude, of
these changes. Coefficients of localization (or variants of them) have
been widely used to compare the geographic distribution of one economic or demographic magnitude to that of another [8, 12]. The
calculation of the coefficient consists of two steps:

1. $A^r/A^{us} - B^r/B^{us} = S^r$ $(r = 1, \ldots, n)$;
2. $\sum_r S^r = CL$ for $S^r > 0$ (or < 0);

where

A = activity A;
B = activity B;
CL = the coefficient of localization; and the superscripts r and
us refer, respectively, to the region and the nation.

The limits of the coefficient are zero and one. If the two activities
are distributed exactly the same, the coefficient will equal the lower
limit; and if A (or B) is concentrated in one region, the value will
approach unity.

For this study, comparisons were made from data over the period
1970–1974. Coefficients of localization were calculated to measure
shifts in the distribution of (1) manufacturing activity relative to
markets; and (2) manufacturing activity relative to energy production, where manufacturing is measured in terms of manufacturing
employment, markets by personal income, and energy production by
total Btu's produced.[4]

Manufacturing is clearly market oriented, as evidence by the coefficients' low values of 0.1388 in 1970 and 0.1364 in 1974 (see Table
7–3). Though not particularly surprising in light of the results of
earlier studies (see note 1), it is interesting that an increase in market
orientation is observed over the period of dramatically rising energy
prices. Further, the coefficients indicate a rather unequal distribution
of manufacturing activity relative to energy production; however,
over the period 1970–1974, these activities have also exhibited evidence of becoming more similarly distributed.

In both cases, empirical observations conform to *a priori* expecta-
tions based on traditional locational analysis. Manufacturing is pre-

Table 7-3. Coefficients of Localization, 1970 and 1974

Coefficients for:	1970	1974	Change
Manufacturing relative to Markets[a]	0.1388	0.1364	-0.0024
Energy Production relative to Manufacturing[b,c]	0.6601	0.6532	-0.0069

[a]U.S. Department of Commerce, *Survey of Current Business*, August, 1971, 1975; Bureau of Labor Statistics, *Handbook of Labor Statistics*, 1975.

[b]U.S. Bureau of Mines, *Coal . . . Bituminous and Lignite, Annual*, 1974; U.S. Bureau of Mines, *Petroleum Statement*, January, December, 1975; U.S. Bureau of Mines, *Minerals Yearbook*, 1970, 1973; American Gas Association, *Gas Facts*, 1971, 1975; U.S. Department of Commerce, *Survey of Current Business*, August, 1971, 1975; Bureau of Labor Statistics, *Handbook of Labor Statistics*, 1975.

[c]Conversion Factors:

  Barrel Crude Petroleum (42 gal.)   =   5,800,000 Btu
  Short Ton Bituminous Coal   = 26,200,000 Btu
  1,000 cu. ft. Natural Gas   =   1,035,000 Btu

dominantly market-oriented and apparently becoming more so. While manufacturing and energy production are unevenly distributed, there has been some movement toward a more equal distribution. Miernyk has shown that energy-producing (surplus) states have realized substantial gains in value-added due to rising energy prices [14]. These same states are those experiencing the comparatively large increases in income potential. Thus, the data suggest that (1) proximity to markets is a dominant concern in industrial location decisions; and (2) energy-producing (surplus) states are beginning to exert themselves as markets. Thus, energy-producing states should attract more manufacturing activity. This is supported by the fact that each of the nine surplus states' share of national manufacturing employment either increased or remained the same over the period 1970 to 1974.

Several limitations of the above procedure should be pointed out. Data availability limited the analysis to the state level. In computing income potentials, the distance variable's exponent was assumed equal to unity. However, since interstate comparisons over time were desired, this weighting problem was ignored. The coefficient of localization is not invariant with the level of aggregation. Moreover, the coefficient can only reveal certain statistical tendencies and cannot identify cause-and-effect relationships.

CONCLUDING REMARKS

The twentieth century has seen an ever-increasing role of the market as a determinant of industrial location. Ullman has demonstrated this graphically by documenting the coincidence of population and economic activity in the industrial heartland of the United States (roughly the New England, Middle Atlantic, and Great Lakes regions) [18]. Additionally, Perloff and Wingo have shown that the tendency for industry to concentrate at the market has yet another dimension [16]. In their discussion of natural resource endowment and economic growth, data are presented that indicate a tendency for fabricating industries to concentrate in the industrial heartland, while first-stage resource users, or processing industries, tend to dominate in the hinterland. Generally, the heartland is diversified, while the hinterland is specialized.

Ullman's findings and those of Perloff and Wingo highlight an historical trend towards a spatial dichotomy of production between heartland and hinterland. However, the bulk of the nation's energy resources are located in the hinterland (Figure 7–4) and rising energy prices have begun to affect the distribution of regional purchasing power. Energy-induced changes in the structure of transport rates, such as those described above, will provide additional incentive to serve this developing market more directly by locating in its proximity. These changes will extend the efficient range of raw material shipments and shorten that of finished goods. Both of these factors will draw an increasing proportion of fabricating industries to the hinterland. In the past, the hinterland's development has been comparatively slow, due, in part, to the relatively weak linkages of dominant industries. However, with the predicted influx of fabricating industries and their effect on the economic structure of energy-producing regions, these areas will become better equipped to promote further economic expansion.

NOTES TO CHAPTER 7

1. See Chinitz and Vernon [7] for a discussion of empirical findings in support of the hypothesis that the pattern of industrial location has tended to conform to that of population. These findings are summarized by Burrows, Metcalf, and Kaler [5, pp. 4–5]. For additional evidence, see Ullman [18] and Harris [9].

2. In support of these arguments, see Association of American Railroads [3] p. 26, and American Waterways Operators, Inc. [2], p. 5., for major commodities shipped by these media. Data from the *U.S. Census of Transportation* on the shipment of manufactured goods confirms their relative dependence on truck transport.

3. $d_{ij} = 1$ when $i = j$.

4. Coal, crude petroleum, and natural gas.

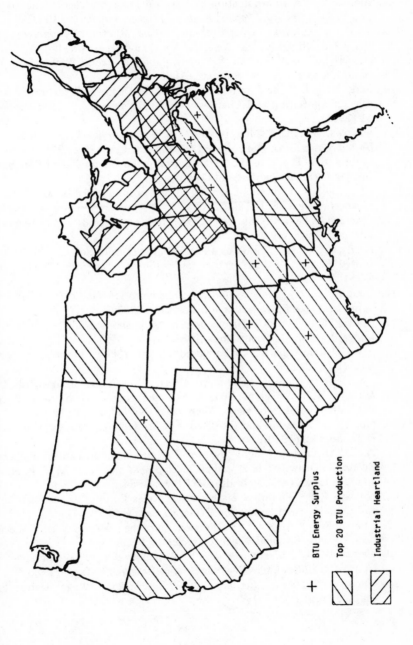

Figure 7—4. Industrial Heartland in Relation to Energy Production

BTU Energy Surplus

Top 20 BTU Production

Industrial Heartland

REFERENCES

1. Alonso, W. "A Reformulation of Classical Location Theory and Its Relation to Rent Theory." *Papers, Regional Science Association* 19 (1967).
2. American Waterways Operators, Inc. *Inland Waterborne Commerce Statistics, 1973.* Arlington, VA: American Waterways Operators, Inc., 1974.
3. Association of American Railroads. *Railroad Facts.* Washington, D.C.: Association of American Railroads, 1975.
4. Barloon, M.J. "The Interrelationship of the Changing Structure of American Transportation and Changes in Industrial Location." *Land Economics* 41 (1965).
5. Burrows, J.C., C.E. Metcalf, and J.B. Kaler. *Industrial Location in the United States.* Lexington, Mass.: D.C. Heath & Company, 1971.
6. Chinitz, B. "The Effect of Transportation Forms on Regional Economic Growth." *Traffic Quarterly* 14 (1960).
7. _____ , and R. Vernon. "Changing Forces in Industrial Location." *Harvard Business Review* 38: 1 (1960).
8. Florence, P. Sargent. *Investment, Location and Size of Plant.* Cambridge, England: University Press, 1948.
9. Harris, C.D. "The Market as a Factor in the Localization of Industry in the United States." *Annals of the Association of American Geographers* 44 (1954).
10. Hirst, E. *Energy Consumption for Transportation in the U.S.* Oak Ridge, Tennessee: Oak Ridge National Laboratory, 1937.
11. Hoover, E.M. *Location Theory and the Shoe and Leather Industries.* Cambridge: Harvard University Press, 1937.
12. _____ . "Redistribution of Population, 1850–1890." *Journal of Economic History* 1 (1941).
13. _____ . *The Location of Economic Activity.* New York: McGraw-Hill, 1948.
14. Miernyk, W.H. "Regional Economic Consequences of High Energy Prices in the United States." *Journal of Energy and Development* 1 (1975).
15. Moses, L. "Location and the Theory of Production." *Quarterly Journal of Economics* 72 (1958).
16. Perloff, H., and L. Wingo. "Natural Resources Endowments and Regional Economic Growth." In *Regional Development and Planning*, J. Friedmann and W. Alonso, eds. Cambridge: MIT Press, 1964.
17. Stewart, J.Q. "Empirical Mathematical Rules Concerning the Distribution and Equilibrium of Population." *Georgia Review* 37 (1947).
18. Ullman, E.L. "Regional Development and the Geography of Concentration." *Papers, Regional Science Association* 4 (1958).
19. Warntz, William. *Macrogeography and Income Fronts.* Monograph Series Number 3. Philadelphia: Regional Science Research Institute, 1965.
20. Weber, A. *Theory of the Location of Industries.* Chicago: University of Chicago Press, 1929.

Bibliography

BOOKS

Abelson, Philip H., ed. *Energy: Use Conservation and Supply.* Washington, D.C.: American Association for the Advancement of Science, 1974.

Adelman, M.A. *The World Petroleum Market.* London and Baltimore: Johns Hopkins University Press, 1972.

Alumni Association of MIT. *The New Wave in the Earth Sciences.* Cambridge: Alumni Association of the Massachusetts Institute of Technology, 1975.

American Petroleum Institute. *A Critical National Choice: New Energy Horizons . . . or Horizontal Disintegration.* Washington, D.C.: American Petroleum Institute, June 1976.

Anderson, Kent P. *A Simulation Analysis of U.S. Energy Demand, Supply, and Prices.* Santa Monica, California: Rand Corporation, October 1975.

_____ , and De Haven, James C. *The Long-Run Marginal Costs of Energy.* Santa Monica, California: Rand Corporation, February 1975.

Anderson, Stephen, et al. *California Energy: The Economic Factors.* San Francisco: Federal Reserve Bank of San Francisco, 1976.

Berlin, Charles J.; Cicchetti, Charles J.; and Gillen, William J. *Perspective on Power: A Study of the Regulation and Pricing of Electric Power.* A Report to the Energy Policy Project of the Ford Foundation. Cambridge, Massachusetts: Lippincott, Ballinger, 1974.

Boesch, Donald F.; Hershner, Carl H.; and Milgram, Jerome H. *Oil Spills and the Marine Environment.* A Report to the Energy Policy Project of the Ford Foundation. Cambridge, Massachusetts: Lippincott, Ballinger, 1974.

Brannon, Bernard M. *Energy Taxes and Subsidies.* A Report to the Energy Policy Project of the Ford Foundation. Cambridge, Massachusetts: Lippincott, Ballinger, 1974.

Brinegar, Claude S. *Oil Company Divesture and the Business Community; or the Senate is Coming!* Los Angeles: Union Oil Company of California, 1976.

Brown, Lester R. *In the Human Interest.* New York: W.W. Norton, 1974.

Committee for Economic Development. *Achieving Energy Independence.* New York: Committee for Economic Development, 1974.

_____ . Research and Policy Committee. *Fighting Inflation and Promoting Growth.* New York: Committee for Economic Development, 1976.

_____ . *International Economic Consequences of High-Priced Energy.* New York: Committee for Economic Development, September 1975.

_____ . *Nuclear Energy and National Security.* New York: Committee for Economic Development, September 1976.

Commoner, Barry. *The Closing Circle.* New York: Alfred A. Knopf, 1971.

_____ . *The Poverty of Power.* New York: Alfred A. Knopf, 1976.

Conference Board. *Energy Consumption in Manufacturing.* A Report to the Energy Policy Project of the Ford Foundation. Cambridge, Massachusetts: Lippincott, Ballinger, 1974.

Exxon Company, U.S.A. *Competition in the Oil Industry.* Houston, Texas: Exxon Company, U.S.A., 1975.

Darmstader, Joel. *An Energy Library.* Washington, D.C.: Resources for the Future, March 1976.

Duchesneau, Thomas D. *Competition in the U.S. Energy Industry.* A Report to the Energy Policy Project of the Ford Foundation. Cambridge, Massachusetts: Lippincott, Ballinger, 1975.

Ehrlich, Paul R., and Ehrlich, Anne H. *The End of Affluence.* New York: Ballantine Books, 1974.

Erickson, Edward W., and Waverman, Leonard, eds. *The Energy Question: An International Failure of Policy.* 2 vols. Toronto and Buffalo: University of Toronto Press, 1974.

Eppen, Gary, ed. *Energy: The Policy Issues.* Chicago: University of Chicago Press, 1975.

Federal Reserve Bank of Boston. *Proceedings of a Conference on New England and the Energy Crisis.* Edgartown, Massachusetts: Federal Reserve Bank of Boston, October 1975.

Ford Foundation. *Exploring Energy Choices.* A Preliminary Report of the Ford Foundation's Energy Policy Project. Washington, D.C.: Ford Foundation, 1974.

_____ , Energy Policy Project. *A Time to Choose: America's Energy Future.* Final Report by the Energy Policy Project of the Ford Foundation. Cambridge, Massachusetts: Lippincott, Ballinger, 1974.

Foster Associates. *Energy Prices 1960–73.* A Report to the Energy Policy Project of the Ford Foundation. Cambridge, Massachusetts: Lippincott, Ballinger, 1974.

Fried, Edward R., et al. *Higher Oil Prices and the World Economy.* Washington, D.C.: The Brookings Institution, 1975.

Georgescu-Roegen, Nicholas. *Energy and Economic Myths: Institutional and Analytical Economic Essays.* New York: Pergamon Press, 1976.

_____ . *The Entropy Law and the Economic Process.* Cambridge, Massachusetts: Harvard University Press, 1971.

Gordon, Richard L. *U.S. Coal and the Electric Power Industry.* Published for Resources for the Future, Inc. Baltimore: Johns Hopkins University Press, 1975.

Gray, John. *Energy Policy: Industry Perspectives.* Cambridge, Massachusetts: Lippincott, Ballinger, 1975.

Gyftopoulos, Elias P.; Lazaridis, Lazaros J.; and Widmer, Thomas F. *Potential Fuel Effectiveness in Industry.* A Report to the Energy Policy Project of the Ford Foundation. Cambridge, Massachusetts: Lippincott, Ballinger, 1974.

Haas, Jerome E.; Mitchell, Edward J.; and Stone, Bernell K. *Financing the Energy Industry.* A Report to the Energy Policy Project of the Ford Foundation. Cambridge, Massachusetts: Lippincott, Ballinger, 1974.

Heilbroner, Robert L. *An Inquiry into the Human Prospect.* New York: W.W. Norton, 1974.

Hollander, Jack M., ed. *Annual Review of Energy.* Vol. 1. Palo Alto, California: Annual Reviews, Inc., 1976.

Holmes, Jay, ed. *Energy, Environment, Productivity.* Proceedings of the First Symposium on RANN. Washington, D.C.: National Science Foundation, 1973.

Institute for Contemporary Studies. *No Time to Confuse.* San Francisco: Institute for Contemporary Studies, 1975.

Kahn, Herman; Brown, William; and Martel, Leon. *The Next 200 Years.* New York: William Morrow, 1976.

Mancke, Richard B. *The Failure of U.S. Energy Policy.* New York: Columbia University Press, 1974.

Massachusetts Institute of Technology. Department of Earth and Planetary Sciences. *The New Wave in Earth Sciences.* Cambridge, Massachusetts: Department of Earth and Planetary Sciences, 1975.

Meadows, Dennis L.; et al. *Dynamics of Growth in a Finite World.* Cambridge: Wright-Allen, 1974.

Medvin, Norman; Lav, Iris J.; and Ruttenberg, Stanley H. *The Energy Cartel: Big Oil vs. The Public Interest.* Washington, D.C.: Marine Engineers' Beneficial Association, 1974/75.

Mendershausen, Horst, and Nehring, Richard. *Protecting the U.S. Petroleum Market against Future Denials of Imports.* A Report Prepared for Defense Advanced Research Projects Agency. Santa Monica: Rand Corporation, October 1974.

Meshan, Ezra J. *The Costs of Economic Growth.* New York: Praeger, 1967.

Meyer, John R. Transportation Solutions to the Energy "Crisis." Supplement to *The Energy Crisis and New Realities.* New York: National Bureau of Economic Research, February 1975.

Miernyk, William H., and Sears, John T. *Air Pollution Abatement and Regional Economic Development.* Lexington, Massachusetts: Lexington Books, D.C. Heath and Company, 1974.

Miller, Roger Leroy. *The Economics of Energy: What Went Wrong?* Glenn Ridge, New Jersey: Thomas Horton and Company, 1974.

National Academy of Sciences, Environmental Studies Board. A Report to the Energy Policies Board of the Ford Foundation. Cambridge, Massachusetts: Lippincott, Ballinger, 1974.

National Bureau of Economic Research. *The Energy Crisis and New Realities.* New York: National Bureau of Economic Research, January 1975.

Nicholls, William H. *Southern Tradition and Regional Progress.* Chapel Hill: University of North Carolina Press, 1960.

Odum, Howard W. *Southern Regions of the United States.* Chapel Hill: University of North Carolina Press, 1936.

Olson, Mancur, and Landberg, Hans H., eds. *The No-Growth Society.* New York: W.W. Norton, 1973.

Olson, McKinley C. *Unacceptable Risk: The Nuclear Power Controversy.* New York: Bantam Books, 1976.

Rand, Christopher T. *Making Democracy Safe for Oil.* Boston: Little, Brown, 1975.

Ray, Dixy Lee. *The Nation's Energy Future.* Washington, D.C.: U.S. Government Printing Office, 1973.

Richardson, Harry W. *Economic Aspects of the Energy Crisis.* Lexington: Lexington Books/Saxon House, 1975.

Sampson, Anthony. *The Seven Sisters.* New York: Viking Press, 1975.

Schlottmann, Alan M. *Environmental Regulation and the Allocation of Coal: A Regional Analysis.* Appalachian Resources Project No. 37. Knoxville: University of Tennessee, August 1975.

Schumacher, E.F. *Small is Beautiful: Economics as if People Mattered.* New York: Harper & Row, 1973.

Szulc, Tad. *The Energy Crisis.* New York: Franklin Watts, 1974.

Winger, John G., et al. *Outlook for Energy in the United States to 1985.* New York: Chase Manhattan Bank, Energy Economics Division, June 1972.

ARTICLES

Abelson, Philip H. "Delays in Tapping Energy Sources." *Science* 187 (10 January 1975): 17.

_____ . "A Global Rush toward Nuclear Energy." *Science* 191 (5 March 1976): 901.

_____ . "Oil and the World's Future." *Science* 194 (12 November 1976): 1.

_____ , and Hammond, Allen L. "The New World of Materials." *Science* 191 (20 February 1976): 633–36.

Adelman, Frank L. "Review of *The Entropy Law and the Economic Process*, by Nicholas Georgescu-Roegen." *Journal of Economic Literature* 10 (June 1972): 458–60.

Adelman, M.A. "Politics, Economics, and World Oil." *American Economic Review* 64 (May 1974): 58–67.

_____ . "Population Growth and Oil Resources." *Quarterly Journal of Economics* 89 (May 1975): 271–75.

_____ . "The Strengths of OPEC." *Resources* 52 (Summer 1976): 7.

_____ , et al. "Energy Self-Sufficiency: An Economic Evaluation." *Technology Review* 76 (May 1974): 23–58.

Alexander, Tom. "Washington's New Glamour Agency Has Some Fancy Plans to Wean Us Off Oil. The Question Is, What Happens to the Private Economy in the Process?" *Fortune* (July 1976): 153–62.

Aliber, Robert Z. "Oil and the Money Crunch." *National Westminster Bank Quarterly Review* (February 1975): 7–19.

Atchison, Sandra. "Geothermal Power: Strangled by Red Tape." *Business Week*, 11 August 1975, pp. 68–69.

Atwood, Genevieve. "The Strip-Mining of Western Coal." *Scientific American* 233 (December 1975): 23–29.

"Back on a Dangerous Binge." *Time*, 30 August 1976, pp. 60–61.

Bennethum, Gary, and Lee, L. Courtland. "Is Our Account Overdrawn?" *Mining Congress Journal* 61 (September 1975): 33–48.

Berg, Charles A. "Conservation in Industry." *Science* 184 (19 April 1974): 264–70.

Bethe, H.A. "The Necessity of Fission Power." *Scientific American* 234 (January 1976): 21–31.

"Bigger Electric Bills Ahead for Big Business." *Business Week*, 29 November 1975, pp. 55–56.

"Blackmail By Oil." *New Republic*, 20 October 1973.

"Blueprint for the Oil Stockpile." *Business Week*, 15 November 1976, pp. 79–84.

Blum, S.L. "Tapping Resources in Municipal Solid Waste." *Science* 191 (20 February 1976): 669–75.

Boffey, Philip M. "NSF Grantee Does Slow Burn as Coal Study Ignites Flap." *Science* 190 (31 October 1975): 446.

_____ . "Nuclear Foes Fault Scientific American's Editorial Judgment in Publishing a Recent Article by Nobel Laureate Hans Bethe." *Science* 191 (26 March 1976): 1248–49.

Boulding, Kenneth E. "The Importance of Improbable Events." *Technology Review* (February 1976): 55–71.

_____ . "The Social System and the Energy Crisis." *Science* 184 (19 April 1974): 255–57.

"Bowing to OPEC." *New Republic* 173 (19 July 1975): 12–14.

"Breeder Reactor Policy." *Science* 191 (26 March 1976): 1214–15.

Bupp, Irvin C., and Derian, Jean-Claude. "The Breeder Reactor in the U.S.: A New Economic Analysis." *Technology Review* 76 (July/August 1974): 26–36.

Carter, Anne P. "Applications of Input-Output Analysis to Energy Problems." *Science* 183 (April 1974): 325–29.

_____ . "Energy, Environment, and Economic Growth." *Bell Journal of Economics and Management Science* 5 (Autumn 1974): 578–92.

Chancellor, W.J., and Goss, J.R. "Balancing Energy and Food Production, 1975–2000." *Science* 192 (16 April 1976): 213–18.

Chilingar, G.V. "An Optimistic Outlook for World Potential Sources of Energy." *Energy Sources* 2 (Winter 1974): 229–32.

Cockburn, Alexander, and Ridgeway, James. "Energy and the Politicians." *New York Review of Books*, 15 April 1976, pp. 19–25.

Commoner, Barry. "Tolling the Bell for Capitalism." *Business Week*, 31 May 1976, pp. 8–10.

Cook, Earl. "Limits to Exploitation of Non-renewable Resources." *Science* 191 (20 February 1976): 677–82.

Craig, Paul P. "The Liquid-Metal Fast Breeder Reactor." *Energy Systems and Policy* 1 (1975): 203–204.

"Cutting the Cost of Free Energy." *Technology Review* 79 (October/November 1976): 23–24.

Daly, Herman E. "The Economics of the Steady State." *American Economic Review* 64 (May 1974): 15–21.

_____. "Energy Demand Forecasting: Prediction or Planning?" *Journal of the American Institute of Planners* 42 (January 1976): 4–15.

Darmstader, Joel, and Landsberg, Hans H. "The Economic Background of the Oil Crisis." *DAEDALUS/Journal of the American Academy of Arts & Sciences* 104 (Fall 1974): 15–37.

"Does Pollution Control Waste Too Much Energy?" *Business Week*, 29 March 1976, p. 72.

Early, John F. "Effect of the Energy Crisis on Employment." *Monthly Labor Review* 97 (August 1974): 8–16.

"Electric Power: Maneuvering in the Coal and Nuclear Debates." *Resources* 51 (Winter 1976): 9–10.

"Energy." *Science* 184 (19 April 1974).

"Energy: The Loss of Innocence." *Resources* 51 (Winter 1976): 1–3.

"Energy by Rail: How Many Cars for Coal?" *Railway Age* 176 (26 May 1975): 28–29.

"Energy Conservation: Congress Acts on Building Standards." *Science* 193 (27 August 1976): 748–53.

"Energy Conservation and Credibility (Letters to the Editor)." *Science* 192 (25 June 1976): 1286.

"Energy Diplomacy." *Science* 192 (30 April 1976): 1.

"Energy Efficiency, European Style." *Technology Review* 79 (October/November 1976): 24.

"Energy Policy: Still No Concensus." *Morgan Guaranty Survey*, June 1975, pp. 4–9.

"Energy Roundup." *Business Week*, 27 July 1974, p. 37.

"Energy Self-Sufficiency: An Economic Evaluation." *Technology Review* 76 (May 1974): 23–58.

"ERDA's Priorities Draw Heavy Fire." *Business Week*, 5 July 1976, pp. 58–59.

"European Breeders (III): Fuels & Fuel Cycle Are Keys to Economy." *Science* 191 (13 February 1976): 551–53.

Faltermayer, Edmund. "Solar Energy Is Here, But It's Not Yet Utopia." *Fortune* (February 1976): 103–106.

Finch, W.I., et al. "Discussion of Uranium Availability and the Breeder Decision." *Energy Systems and Policy* 1 (1975): 259–70.

First National City Bank. "How Prices Will Foil the New Malthusians." *Monthly Economic Letter*, December 1975, pp. 6–9.

"The Flaky Arguments over Breaking Up Big Oil." *Business Week*, 16 August 1976, pp. 93–98.

"Flash Hydrogenation of a Bituminous Coal." *Science* 189 (5 September 1976): 793–95.

"A Flood of Foreign LNG for U.S. Factories." *Business Week*, 13 October 1975, pp. 101–104.

"Food and Agriculture." *Scientific American* 235 (September 1976): 30–39.

"Forced Divesture in Oil?" *Morgan Guaranty Survey*, June 1976, pp. 3–10.

Ford, Andrew. "Environmental Policies for Electricity Generation: A Study of the Long-Term Dynamics of the SO_2 Problem." *Energy Systems and Policy* 1 (1975): 287–304.

"Foundations of the Model of Doom." *Science* 189 (26 September 1975): 1077–78.

Fried, Edward R. "International Trade in Raw Materials: Myths and Realities." *Science* 191 (20 February 1976): 641–46.

"Fusion Research (I): What is the Program Buying the Country?" *Science* 192 (25 June 1976): 1320–23.

"Fusion Research (II): Detailed Reactor Studies Identify More Problems." *Science* 193 (2 July 1976): 38–40.

"Geological Survey Lowers Its Sights." *Science* 189 (18 July 1975): 200.

Georgescu-Roegen, Nicholas. "Economic Growth and Its Representation by Models." *Atlantic Economic Journal* 4 (Winter 1976): 1–8.

_____ . "Energy and Economic Myths." *Southern Economic Journal* 41 (January 1975): 347–81.

_____ . "The Entropy Law and the Economic Problem." Paper presented at Distinguished Lecture Series No. 1 at the University of Alabama, 3 December 1970.

_____ . "Mechanistic Dogma and Economics." *Methodology and Science* 7 (1974): 174–83.

_____ . "Richard T. Ely Lecture: The Economics of Production." *American Economic Review* 60 (May 1970): 1–9.

"Geothermal Power: Strangled by Red Tape." *Business Week*, 11 August 1975, pp. 68–69.

"A Giant Step for Solar Heating." *Business Week*, 18 October 1976, pp. 99–100.

Giarratani, Frank. "Application of an Interindustry Supply Model to Energy Issues." *Environment and Planning* 8 (June 1976): 447–54.

_____ , and Socher, Charles F. "The Pattern of Industrial Location and Rising Energy Prices." *Atlantic Economic Journal* 5 (March 1977): 48–55.

Gillette, Robert. "Geological Survey Lowers Its Sights." *Science* 189 (18 July 1975): 200.

_____ . "Oil and Gas Resources: Academy Calls USGS Math 'Misleading.' " *Science* 187 (28 February 1975): 723–27.

Goeller, H.E., and Weinberg, Alvin M. "The Age of Substitutability." *Science* 191 (20 February 1976): 683–89.

Gordon, Lincoln. "Limits to the Growth Debate." *Resources* 52 (Summer 1976).

"The Great Nuclear Debate." *Time*, 8 December 1975, pp. 36–41.

Gregory, Derek P. "The Hydrogen Economy." *Scientific American* 228 (January 1973): 13–21.

Griffin, James M. "The Effects of Higher Prices on Electricity Consumption." *Bell Journal of Economics and Management Science* 5 (Autumn 1974): 515–39.

Hamester, Hans L.; Graves, Glen A.; and Plummer, James L. "Risk Aversion and Energy Policy: A Case for Breeder Research and Development." *Energy Systems and Policy* 1 (1975): 233–58.

Hammond, Allen L. "Coal Liquefaction Plant Goes Ahead." *Science* 194 (12 November 1976): 712.

Hausman, Jerry A. "Project Independence Report: An Appraisal of U.S. Energy Needs up to 1985." *Bell Journal of Economics* 6 (Autumn 1975): 517–51.

Hayes, Earl T. "Energy Implications of Materials Processing." *Science* 191 (20 February 1976): 661–65.

Hirst, Eric. "Transportation Energy Conservation Policies." *Science* 192 (2 April 1976): 15–20.

Holdren, John P. "Uranium Availability and the Breeder Decision." *Energy Systems and Policy* 1 (1975): 205–32.

"How Much to Pay the OPEC Piper?" *Time*, 8 November 1976, p. 92.

Huddle, Franklin P. "The Evolving National Policy for Materials." *Science* 191 (20 February 1976): 654–59.

Hudson, E. A., and Jorgenson, D. W. "U.S. Energy Policy and Economic Growth, 1975–2000." *Bell Journal of Economics and Management Science* 5 (Autumn 1974): 461–514.

"It's a Long Haul from Mine to Reactor." *Fortune* (October 1975): 153–57.

Kennedy, Michael. "An Economic Model of the World Oil Market." *Bell Journal of Economics and Management Science* 5 (Autumn 1974): 540–77.

Keyfritz, Nathan. "World Resources and the World Middle Class." *Scientific American* 235 (July 1976): 28–35.

"King Coal's Return: Wealth and Worry." *Time*, 1 March 1976, pp. 45–47.

Kraar, Louis. "OPEC Is Starting to Feel the Pressure." *Fortune* (May 1975): 186–91.

Kutscher, Ronald E., and Bowman, Charles T. "Industrial Use of Petroleum: Effect on Employment." *Monthly Labor Review* 97 (March 1974): 3–8.

"The Lag in Coal Expansion." *Business Week*, 27 January 1975, pp. 127–30.

Lambert, Richard D., and Heston, Alan W., eds. *Annals of the American Academy of Political and Social Science—Adjusting to Scarcity* 420 (July 1975).

Landsberg, Hans H. "Materials: Some Recent Trends and Issues." *Science* 191 (20 February 1976): 637–41.

Lapp, Ralph E. "We May Find Ourselves Short of Uranium, Too." *Fortune* (October 1975): 151–52.

Leontief, Wassily. "Environmental Repercussions and the Economic Structure: An Input-Output Approach." *Review of Economics and Statistics* (August 1970): 262–70.

Lewin, Nathan. "Justice for the Arab Boycott." *New Republic*, 6 September 1975, pp. 13–16.

Lieberman, M. A. "U.S. Uranium Resources—An Analysis of Historical Data." *Science* 192 (30 April 1976): 431–36.

"Limits to Growth '75: Long-Distance View of a Mad Hatter's Tea Party." *Technology Review* 78 (January 1976): 19–20.

"Limits to Growth: Texas Conference Finds None, But Didn't Look Too Hard." *Science* 190 (7 November 1975): 540–41.

Lincoln, G.A. "Energy Conservation." *Science* 180 (13 April 1973):155–80.

MacGregor, Ian D. "Natural Distribution of Metals and Some Economic Effects." *Annals of the American Academy of Political and Social Science* 420 (July 1975): 31–45.

McKeon, Mark. "High Tension on the Prairie." *Nation*, 30 October 1976, pp. 434–36.

McKetta, John. "I Predict . . . A Hair-Curling Energy Crisis by 1985." *Electric Perspectives*, January 1976.

McKie, James W. "Review of *The Energy Question: An International Failure of Policy*, by Edward W. Erickson and Leonard Waverman, editors." *Bell Journal of Economics* 6 (Autumn 1975): 720–23.

_____. "The Political Economy of World Petroleum." *American Economic Review* 64 (May 1974): 51–57.

Manne, Alan S. "Waiting for the Breeder." Paper presented at the Sixth International Conference on Input-Output Techniques, Vienna, 22–26 April 1974.

Maugh, Thomas H., II. "Natural Gas: United States Has It if the Price Is Right." *Science* 191 (13 February 1976): 549–50.

Meredith, Dennis L. "A Nuclear New England . . . Unavoidably Economical." *Technology Review* 78 (January 1976): 18–19.

Metz, William D. "Basic Research Funding: ERDA De-energizes Nuclear Science." *Science* 191 (5 March 1976): 931–33.

_____. "European Breeders (III): Fuels and Fuel Cycle Are Keys to Economy." *Science* 191 (13 February 1976): 551–53.

Miernyk, William H. "The Changing Structure of the Southern Economy." Southern Growth Policies Board Occasional Paper, No. 2, January 1977.

_____. "Coal and the Future of the Appalachian Economy." *Appalachia* 9 (October/November 1975): 29–35.

_____. "Decline of the Northern Perimeter." *Society* 13 (May/June 1976): 24–26.

_____. "Regional Employment Impacts of Rising Energy Prices." *Labor Law Journal* 26 (August 1975): 518–23.

"The MIT Report: Is Doomsday Really That Close?" *Business Week*, 11 March 1972, pp. 97–98.

Murphy, Edward E., and Perez-Lopez, Jorge F. "Trends in U.S. Export Prices and OPEC Oil Prices." *Monthly Labor Review* (November 1975): 36–43.

Naill, Roger F.; Meadows, Dennis L.; and Stanley-Miller, John. "The Transition to Coal." *Technology Review* 78 (October/November 1975): 2–12.

"The Next Rise in Oil Prices." *Business Week*, 31 May 1976, pp. 20–22.

Nordhaus, William D. "Resources as a Constraint on Growth." *American Economic Review* 64 (May 1974): 22–26.

"Nuclear Foes Fault Scientific American's Editorial Judgement in Publishing a Recent Article by Nobel Laureate Hans Bethe." *Science* 191 (26 March 1976): 1248–49.

"An OPEC Ploy to Raise Oil Prices?" *Business Week*, 5 July 1976, pp. 21–22.

"Panel Calls Beneficiation—FGD Combination 'Most Economical Best All-

Around Choice.' " *Journal of the Air Pollution Control Association* 25 (November 1975): 1115—18.

Parisi, Anthony. "The Pressure on OPEC to Raise Oil Prices Again." *Business Week*, 26 May 1975, pp. 28.

Peck, A.E., and Doering, O.C., III. "Voluntarism and Price Response: Consumer Reaction to the Energy Shortage." *Bell Journal of Economics* 7 (Spring 1976): 287—92.

Perry, Harry. "The Gasification of Coal." *Scientific American* 230 (March 1974): 19—25.

"Petroleum Prices and the New Energy." *Resources* 51 (Winter 1976): 4.

"Pipelining Coal." *Business Week*, 22 December 1975, p. 56.

"Plutonium and Christian Ethics (Letters to the Editor)." *Science* 192 (21 May 1976): 738, 740.

Posner, Michael. "Energy at the Center of the Stage." *Three Banks Review* 104 (December 1974): 3—27.

"Project Independence: A Critical Look." *Chemical Engineering* 82 (6 January 1975): 92—105.

Rand, Christopher. "Making Democracy Safe for Oil." *New Republic*, 15 September 1975, pp. 26—28.

Rearden, W.A. "Input-Output Analysis of U.S. Energy Consumption." Paper presented at the Sixth International Conference on Input-Output Techniques, Vienna, 22—26 April 1974.

"The Riddle of the Gas Shortage." *Resources* 51 (Winter 1976): 5.

Roberts, Marc J. "Review of *The World Petroleum Market*, by M.A. Adelman." *Journal of Economic Literature* 12 (December 1974): 1363—68.

Rolfe, Sidney E. "Whatever Happened to Project Independence?" *Saturday Review*, 25 January 1975, pp. 25—28.

Rose, David J. "Commentary on the Foregoing Breeder Papers and on the Problem in General." *Energy Systems and Policy* 1 (1975): 271—76.

Rose, Sanford. "Why Big Oil is Putting the Brakes On." *Fortune* 93 (March 1976): 110—15, 173—76.

Roy, Rustum. "Energy Education." *Science Education News*, May 1974.

Russell, Milton. "Turning Point in Oil Policy." *Resources* 52 (Summer 1976): 7.

Samuelson, Robert. "The Oil Policy We Needed But Didn't Get—Too Little Too Late." *New Republic*, January 1975, pp. 12—19.

"Saving Two Decades on a Fusion Technique." *Business Week*, 15 November 1976, pp. 125—27.

Schachter, Gustav. "Review of *The Kingdom of Oil: The Middle East, Its People and Its Power*, by Ray Vicker." *Growth and Change* 6 (July 1975): 57—58.

Schipper, Lee. "Energy Use in Sweden: Some Lessons for Americans." *Current Sweden*, November 1976, pp. 1—5.

Schneider, Alan M. "Elasticity of Demand for Gasoline." *Energy Systems and Policy* 1 (1975): 277—86.

Schultze, Charles L. "Another Oil Recession." *New Republic*, 5 & 12 July 1975, pp. 8—16.

"The Scramble for Resources." *Business Week*, 30 June 1973, pp. 56—63.

"The Second War Between the States." *Business Week*, 17 May 1976, pp. 92—114.

Shapley, Deborah. "Senate Study Predicts U.S. Oil 'Exhaustion.' " *Science* 187 (21 March 1972): 1064.

Shepherd, Stephen B. "How Much Energy Does the U.S. Need?" *Business Week*, 1 June 1974, pp. 69—70.

Solow, Robert M. "The Economics of Resources or the Resources of Economics." *American Economic Review* 64 (May 1974): 1—14.

Spaid, Ora. "Forecast: Doubled Coal Production in Appalachia." *Appalachia* (June/July 1975): 1—10.

"The Squeeze on the Nation's Public Power." *Business Week*, 12 April 1976, pp. 73—74.

Squires, Arthur M. "Chemicals from Coal." *Science* 191 (20 February 1976): 689—99.

_____. "Clean Power from Dirty Fuels." *Scientific American* 227 (October 1972): 26—35.

Starr, Chauncey. "Energy and Power." *Scientific American* 224 (September 1971): 36—49.

Steinhart, John S., and Steinhart, Carol E. "Energy Use in the U.S. Food System." *Science* 184 (19 April 1974): 307—16.

"Trying to Scuttle a Coal Pipeline." *Business Week*, 15 December 1975, p. 15.

Ulmer, Melville J. "Thwarting the Cartel." *New Republic*, 15 February 1975, pp. 9—10.

"The Uranium Dilemma: Why Prices Mushroom." *Business Week*, 1 November 1976, pp. 92—97.

"U.S. Recovery: Not Yet Complete." *Business in Brief*, April 1976.

"Utilities: Weak Point in the Energy Future." *Business Week*, 20 January 1975, pp. 46—54.

van de Walle, Etienne. "Review of *Dynamics of Growth in a Finite World*, by Dennis L. Meadows, William W. Behrens, III, Donella H. Meadows, Roger F. Naill, Jorgen Randers, and Erich K. O. Zahn." *Science* 189 (26 September 1975):1077—78.

Wade, Nicholas. "Nicholas Georgescu-Roegen: Entropy is the Measure of Economic Man." *Science* 190 (31 October 1975): 447—50.

Waelti, John J. "Review of *Energy: The New Era*, by S. David Freeman." *Growth and Change* 6 (July 1975): 57—58.

"Awash with Oil." *New Republic*, 15 May 1976, pp. 3—4.

Weinstein, Milton C., and Zeckhauser, Richard J. "The Optimal Consumption of Depletable Natural Resources." *Quarterly Journal of Economics* 89 (August 1975): 371—92.

"Why Atomic Power Dims Today." *Business Week*, 17 November 1975, pp. 98—105.

"Why OPEC's Rocket Will Lose Its Thrust." *First National City Bank Monthly Economic Letter*, June 1975, pp. 11—15.

"Why the Russians Go All-Out for Nuclear Power." *Business Week*, 2 August 1976, pp. 52—53.

Yergin, Daniel. "How Europe Saves Energy." *New Republic*, 13 December 1975, pp. 14–21.

_____ . "OPEC Imperium." *New Republic*, 15 November 1975, pp. 15–18.

Zebroski, E.L., and Minnick, L.E. "Breeder Reactor Policy." *Science* 191 (26 March 1976): 1214–15.

DOCUMENTS

Huntington, Hillard, and Kahn, James. "Regional Industrial Growth and the Price of Energy." Working Paper 76–WPA–35. Washington, D.C.: Federal Energy Administration, May 1976.

Polenske, Karen R., and Levy, Paul F. *Multiregional Economic Impacts of Energy and Transportation Policies.* DOT Report No. 8. Prepared for the University Research Program, U.S. Department of Transportation. Washington, D.C.: Government Printing Office, March 1975.

U.S. Congress. Joint Economic Committee. *Achieving the Goals of the Employment Act of 1946—Thirtieth Anniversary Review.* Joint Committee Print, Volume 2—Energy, Paper 1. Washington, D.C.: Government Printing Office, 1975.

_____ . *Adequacy of U.S. Oil and Gas Reserves. Hearings before the Joint Economic Committee.* 94th Cong., 1st sess., 25 February 1975.

_____ . *Changing Conditions in the Market for State and Local Government Debt.* Joint Committee Print. Washington, D.C.: Government Printing Office, 1976.

_____ . *The Current Fiscal Position of State and Local Governments.* 94th Cong., 1st sess., 17 December 1975.

_____ . *The Economic Impact of Forthcoming OPEC Price Rise and "Old" Oil Decontrol. Hearings before the subcommittee on Consumer Economics.* 94th Cong., 1st sess., 1975.

_____ . "Electricity Generation Using Solar Energy Examined in JEC Hearing." *Notes from the Joint Economic Committee*, 8 May 1976.

_____ . *The Fast Breeder Reactor Decision: An Analysis of Limits and the Limits of Analysis.* Joint Committee Print. Washington, D.C.: Government Printing Office, 1976.

_____ . *Horizontal Integration of the Energy Industry. Hearings before a subcommittee on Energy.* 94th Cong., 1st sess., 1975.

_____ . "JEC Energy Subcommittee Explores Means to Reduce Our Energy Consumption." *Notes from the Joint Economic Committee*, 14 July 1976, p. 1.

_____ . "Multi-national Oil Companies and OPEC Implications for U.S. Policy." *Notes from the Joint Economic Committee*, 13 July 1976, pp. 1–3.

_____ . *Reappraisal of Project Independence Blueprint. Hearings before the Joint Economic Committee.* 94th Cong., 1st sess., 18 March 1975.

_____ . *A Review and Update of the Cost-Benefit Analysis for the Liquid Metal Fast Breeder Reactor.* Joint Committee Print. Washington, D.C.: Government Printing Office, 1976.

_____ . *State and Local Government Credit Problems. Hearings before the Joint Economic Committee.* 94th Cong., 1st sess., 1975.

_____ . *Technology and Economic Growth. Hearings before the subcommittee on Economic Growth.* 94th Cong., 1st sess., 1975.

_____. *Technology, Economic Growth, and International Competitiveness.* Joint Committee Print prepared by Robert Gilpin. Washington, D.C.: Government Printing Office, 9 July 1975.

_____. *U.S. Foreign Energy Policy. Hearings before a subcommittee on Energy.* 94th Cong., 1st sess., 1975.

U.S. Congress. Office of Technology Assessment. Committees on Science and Technology, Senate Interior and Insular Affairs, and the Joint Committee on Atomic Energy. *An Analysis Identifying Issues in the Fiscal Year 1976 ERDA Budget.* Joint Committee Print, Serial D. Washington, D.C.: Government Printing Office, 1975.

_____. *An Analysis of the ERDA Plan and Program.* Washington, D.C.: Government Printing Office, October 1975.

_____. *Comparative Analysis of the 1976 ERDA Plan and Program.* Washington, D.C.: Government Printing Office, May 1976.

_____. *Energy, the Economy and Mass Transit.* Washington, D.C.: Government Printing Office, December 1975.

U.S. Congress. Senate. *Cornell Workshop on Energy and the Environment.* S. Res. 45. 92d Cong., 2d sess., May 1972.

_____. Committee on Interior and Insular Affairs. *Energy Policy Papers.* 93d Cong., 2d sess., 1974.

_____. Permanent Subcommittee on Investigations of the Committee on Government Operations. *Staff Study of the Oversight and Efficiency of Executive Agencies with Respect to the Petroleum Industry, Especially as It Relates to Recent Fuel Shortages.* 93d Cong., 1st sess., 8 November 1973.

U.S. Department of the Interior. Bureau of Mines. *Long-Distance Coal Transport: Unit Trains or Slurry Pipelines,* by T.C. Campbell and Sidney Katell. Information Circular 8690. Washington, D.C.: Government Printing Office, 1975.

_____. *United States Energy Through the Year 2000,* by Walter G. Dupree, Jr. and John S. Corsentino. Washington, D.C.: Government Printing Office, December 1975.

U.S. Department of the Interior. Federal Energy Administration. *1976 Executive Summary: National Energy Outlook.* Washington, D.C.: Government Printing Office, 1976.

_____. Council on Environmental Quality. *Project Independence Blueprint: Final Task Force Report.* Volume 2, *Project Independence and Energy Conservation: Transportation Sectors.* Washington, D.C.: Government Printing Office, November 1974.

_____. *Project Independence: Coal.* Washington, D.C.: Government Printing Office, November 1974.

_____. *Project Independence Report.* Washington, D.C.: Government Printing Office, November 1974.

U.S. Department of Labor. "Employment in Nuclear Energy Activities." *News,* 28 June 1976, pp. 1–5.

_____. *The U.S. Economy in 1985.* Bulletin 1809. Washington, D.C.: Government Printing Office, 1974.

U.S. Energy Research and Development Administration. *A National Plan for Energy Research, Development, and Demonstration: Creating Energy Choices for the Future*, Vol. 1, *The Plan* (ERDA−48). Washington, D.C.: Government Printing Office, 1975.

U.S. President. *International Economic Report of the President*. Washington, D.C.: Government Printing Office, March 1975.

NEWSPAPERS

Adams, James Ring. "Bringing OPEC to the Bar." *Wall Street Journal*, 20 May 1975.

"Algeria Apparently Cuts Price of Crude: Arab Producers Urged to Restudy Policy." *Wall Street Journal*, 11 March 1975.

Aliber, Robert Z. "Impending Breakdown of OPEC Cartel." *Wall Street Journal*, 20 March 1975.

Anderson, Jack. "The Almighty Petrodollar and the Arab Boycott." *Washington Post*, 26 September 1976.

Arble, Meade. "Notes from a Coal Mine." *New York Times*, 12 January 1975.

Austin, Danforth W. "Gasoline Usage Rises as Price and Supply Brings End to Frugality." *Wall Street Journal*, 3 May 1976.

Carley, William M. "Fuel Shortage Forecast for U.S. Nuclear Plants within Decade or Two." *Wall Street Journal*, 7 June 1976.

Chace, James. "Review of *Making Democracy Safe for Oil*, by Christopher T. Rand." *New York Times Book Review*, 24 August 1975.

_____ . "Review of *The Great Detente Disaster*, by Edward Friedland, Paul Seabury, and Aaron Wildavsky." *New York Times Book Review*, 24 August 1975.

Cheng, Chu-Yuan. "China's Future as an Oil Exporter." *New York Times*, 4 April 1976.

Church, Frank. "Coping with the Cartel: An Alternative Program." *Washington Post*, 23 February 1975.

Clark, Lindley H., Jr. "Many Analysts See Weakening of OPEC, Oppose Concessions by U.S. to Oil Cartel." *Wall Street Journal*, 4 March 1975.

"Climatologists Say Earth's Crust Shows Coming of Ice Age." *New York Times*, 29 August 1976.

Collins, Joseph. "Glut of 'Gas' Brings British Price War." *New York Times*, 12 October 1975.

Cowan, Edward. "Economics of Nuclear Power Are No Longer So Optimistic." *New York Times*, 18 July 1976.

_____ . "Share of Arab Oil in U.S. Imports Up." *New York Times*, 25 May 1975.

_____ . "Solar Energy Is Gaining as Future Fuel Solution." *New York Times*, 1 June 1975.

_____ . "Soviet Said to Bar Bid by U.S. to Buy Oil at a Discount." *New York Times*, 12 October 1975.

_____ . "U.S. Depends More Than Ever on Imported Oil." *New York Times*, 18 July 1976.

Crittenden, Ann. "Economist Hero Thinks Small." *New York Times*, 26 October 1975.

Darmstader, Joel. "Energy Books." *Washington Post Book World*, 11 January 1976.

de Onis, Juan. "OPEC, Facing Real Losses, May Impose Oil Increases." *Wall Street Journal*, 1 June 1975.

Faber, Harold. "Site for Big Power Plant Sets Off a Debate Upstate." *New York Times*, 25 July 1976.

Fallaci, Oriana. "A Sheik Who Hates to Gamble." *New York Times Magazine*, 14 September 1975.

Fandell, Todd, and Camp, Charles. "Transportation in 2000 to Rely on Equipment Much Like Today's." *Wall Street Journal*, 1 April 1976.

Finney, John W. "Focusing on Energy—Unclearly." *New York Times*, 9 March 1975.

"FPC Sees Electric Bills Tripling by 1990, Cities Growing Delays in Plant Start-Ups." *Wall Street Journal*, 17 April 1972.

Goshko, John M. "OPEC—From Ineptitude to World's Most Powerful Cartel." *Washington Post*, 22 December 1974.

Halloran, Richard. "The Changing Face of the Oil Crisis." *New York Times*, 6 April 1975.

Harris, Louis. "Oil or Israel." *New York Times Magazine*, 6 April 1975.

Hayes, Dennis. "Conservation as a Major Energy Source." *New York Times*, 21 March 1976.

Hill, Gladwin. "Limits on A-Plants Facing Lawsuits." *New York Times*, 29 August 1976.

House, Karen Elliot. "Administration Cuts Back Its Plan to Develop a Plutonium Reactor to Generate Electricity." *Wall Street Journal*, 2 June 1975.

Kalborn, Peter T. "A-Power Opposition Growing in Europe." *New York Times*, 26 September 1976.

Kessler, Felix. "OPEC to Permit Reduction in Oil Prices by Smaller Members in Bid to Avoid Split." *Wall Street Journal*, 27 February 1975.

Kraft, Joseph. "The New Regional Balance." *Washington Post*, 13 June 1976.

Lichtenstein, Grace. "31 in House Ask a Delay on Power Plant." *New York Times*, 11 April 1976.

MacDonald, Stephen. "Review of *The Closing Circle*, by Barry Commoner." *Wall Street Journal*, 22 April 1971.

Martin, Douglas. "Ocean Divers Expand Oil-Gas Search Role, Go Lower and Lower." *Wall Street Journal*, 1 September 1976.

McPherson, Elizabeth. "Review of *Energy for Survival: The Alternative to Extinction*, by David Howell." *Washington Post Book World*, 11 January 1976.

———. "Review of *Producing Your Own Power: How To Make Nature's Energy Sources Work For You*, by Carol Huppingstone." *Washington Post Book World*, 11 January 1976.

Mintz, Morton. "Natural Gas Price Said Higher Than Is Justified." *Washington Post*, 31 October 1976.

"National Iranian Oil Company Advertisement." *Wall Street Journal*, 6 June 1975.

Netschert, Bruce C. "The Cloud on OPEC's Horizon." *Wall Street Journal*, 29 March 1976.

"The Oil Is Synthetic, but the Problems Are Real." *New York Times*, 12 October 1975.

"OPEC Likely to Boost Oil Prices in 1977, but Less than 10%, Sheikh Yamani Says." *Wall Street Journal*, 10 August 1976.

"OPEC States Advised of Global Problems that May Follow Any Oil Price Increase." *Wall Street Journal*, 27 October 1976.

Passell, Peter. "Review of *The Poverty of Power*, by Barry Commoner." *New York Times Book Review*, 23 May 1976.

Ricklefs, Roger. "Cities May Flourish in South and West, Decline in Northeast." *Wall Street Journal*, 6 April 1976.

"Saudi Arabia to Embargo All Oil to U.S." *Washington Post*, 21 October 1973.

"Saudi-Production of Crude Oil Is Cut." *Washington Post*, 9 March 1975.

Semple Robert B., Jr. "Amid Its Gloom, Britain at Least Has the Hope of North Sea Oil." *New York Times*, 31 October 1976.

Sherrill, Robert. "Breaking Up Big Oil." *New York Times Magazine*, 3 October 1976.

Shilling, Gary. "Lessons of History for OPEC." *Wall Street Journal*, 10 March 1975.

Smith, William N. "U.S. Is Further Than Ever from Oil Independence." *New York Times*, 22 August 1976.

Stuart, Reginald. "Rising Electric Bills Drive Consumers to Organize." *New York Times*, 4 April 1976.

"Study Says Energy Use Could Be Cut in Half." *Washington Post*, 1 February 1976.

Tanner, James C. "No Crippling Shortage of Energy Expected, but Cost Will Be High." *Wall Street Journal*, 29 March 1976.

"Tarnishing the Atom . . . Policy Alternative." *New York Times*, 12 February 1976.

"Texas Oil Production May Be Stabilizing after Years of Decline, State Agency Says." *Wall Street Journal*, 21 May 1975.

"To Thwart a Cartel (Letters to the Editor)." *New York Times*, 18 May 1975.

"U.S. Agency Sharply Reduces Estimate of Undiscovered Oil and Gas Reserves." *Wall Street Journal*, 8 May 1975.

"U.S. Oil Use Is Rising as OPEC Gets Ready to Set Higher Prices." *Wall Street Journal*, 11 November 1976.

"Utility's Nuclear Plants Issues Most of its Power." *New York Times*, 29 August 1976.

Vicker, Ray. "OPEC Tests an Uneasy Unity." *Wall Street Journal*, 27 February 1975.

"Wasting Energy . . . and Conserving It." *New York Times*, 29 February 1976.

Wilson, David Gordon. "Letter to the Editor: Cheap-Energy Society." *New York Times*, 23 February 1975.

Winski, Joseph M. "By 2000, Prevention of Starvation May Be Chief Global Concern." *Wall Street Journal*, 25 March 1976.

Yergin, Daniel. "The Economic Political Military Solution." *New York Times Magazine*, 16 February 1975.

Index

About the Authors

Dr. William H. Miernyk is Benedum Professor of Economics and Director of the Regional Research Institute at West Virginia University. He did his undergraduate work at the University of Colorado and received his M.A. and Ph.D. degrees at Harvard University. He taught at Northeastern University and the University of Colorado, and has been a visiting professor of economics at the Massachusetts Institute of Technology and Harvard University.

Dr. Miernyk has served as a consultant to the U.S. Senate Committee on Commerce and the Special Committee on Unemployment Problems. He has been a consultant to the Appalachian Regional Commission and served on advisory committees in the U.S. Departments of Commerce and Labor. He has worked with the common market countries of Central America as a consultant to the Brookings Institution Foreign Policy Project.

Dr. Miernyk has written widely on labor economics, regional economic development, and input-output analysis. He is the author or co-author of eight books, has contributed to seventeen additional books, and has written over 100 articles and papers.

Frank Giarratani is an Assistant Professor of Economics at Georgetown University. From 1973 to 1976 he held the positions of Research Fellow and Research Associate at the Regional Research Institute, West Virginia University. His major area of interest if Regional/Urban Economics with special emphasis on interdisciplinary analysis. His published work in this area includes applications to envi-

ronmental and energy problems as well as the evaluation of bipropor-
tional matrix estimation techniques.

Charles F. Socher is a Ph.D. candidate in economics at West Virginia University. As a member of the Regional Research Institute, his work has focused upon input-output analysis and energy issues. His previous publications include *Regional and Interregional Input-Output Analysis: An Annotated Bibliography* and *The West Virginia Input-Output Model: A User's Guide* (co-author).